Society, Environment and Statistics

This interdisciplinary book series demonstrates the significance of statistical science in quantifying and understanding societal and environmental changes, while effectively managing resources. The series invites manuscripts that cover a wide range of themes, including original research in traditional fields, emerging topics, practical applications, and connections between novel findings and established works. Particularly encouraged are submissions that showcase methodologies, case studies, good practices and innovative solutions addressing the complex challenges of the United Nations' Sustainable Development Goals (SDGs). Rigorously peer-reviewed, all contributions adhere to the highest standards of scientific literature.

Claus Weihs • Walter Krämer • Sarah Buschfeld
Editors

Statistics Today

Everyday Applications, Research Questions,
Insights, and Challenges

 Springer

Editors

Claus Weihs
Department of Statistics
TU Dortmund University
Dortmund, Germany

Walter Krämer
Department of Statistics
TU Dortmund University
Dortmund, Germany

Sarah Buschfeld
Department of Cultural Studies
TU Dortmund University
Dortmund, Germany

ISSN 2948-2763   ISSN 2948-2771   (electronic)
Society, Environment and Statistics
ISBN 978-3-662-68906-6   ISBN 978-3-662-68907-3   (eBook)
https://doi.org/10.1007/978-3-662-68907-3

To Heidrun, Max, Doris, and Leo

Preface

"I keep saying the sexy job in the next ten years will be statisticians ... The ability to take data - to be able to understand it, to process it, to extract value from it, to visualize it, to communicate it is going to be a hugely important skill in the next decades."
(Hal Varian, chief economist at Google, in 2009;[1] more than a decade later, we can see that Varian was right.)

Academic disciplines have their ups and downs. Some, such as medicine for obvious reasons, have experienced a stable interest. Ever since Plato and Aristotle, philosophy has enjoyed the attention of contemporaries of every century, in particular since it has never been well understood. Other disciplines have moved into the focus of general attention only sporadically. The early twentieth century was the high time of chemistry, technology, and physics: artificial fertilizers, airplanes, dyestuffs, electrification, rockets, and nuclear fission were hotly debated by experts and non-experts alike; Konrad Röntgen, the inventor of X-ray technology, was the first scientific pop star ever. Then, in the wake of Siegmund Freud, psychology was the number one party topic in certain social circles for a while; a person who could not tell about his/her last visit to a psychoanalyst was quasi socially declassed. And at the end of the 1960s, sociology was, for a short time, regarded as the key to decode the future of humanity.

Currently, statistics is one of the most popular sciences. Spurred by an enormous revolution in both data processing and data acquisition, new opportunities and challenges arise almost every day. Even the most superficial newspaper reader cannot avoid realizing that data science in general and statistics in particular are central aspects of current science and our everyday lives. Today, everyone talks about 'Big Data' and corresponding degree courses are sprouting like mushrooms. However, it has been often overlooked that the following topics are anything but new: correct sampling and handling of data, decision making under uncertainty, differentiating between chance and deterministic patterns, and extrapolating from samples to the

---

[1] See https://flowingdata.com/2009/02/25/googles-chief-economist-hal-varian-on-statistics-and-data/, visited 11.9.2019.

general. For decades, thousands of researchers in statistics departments, in business, and industry all over the world have been active in these fields. The current major increase in excitement and activity originates in the vast amount of data that require statistical analysis.

The data sources, however, have always existed. Ambient temperature on the northern side of the Matterhorn on Christmas Eve has been ready to be observed ever since the Matterhorn has existed. And that Mrs. X bought two bottles of Chateau Baron on January 10, 1998 at a certain Aldi-shop in Berlin was always a fact. This information was, however, limited to Mrs. X and the person working the cash register. Today, if Mrs. X payed by credit card, the whole world could know. The current 'Big Data' hype is thus the result of the recent availability of the data, for which the sources always existed, but which were simply never exploited. For most of human history, people knew comparatively little. This has changed due to the interaction of increasingly efficient computer and storage technologies and advanced methods and technologies for data exploitation. Therefore, large amounts of data are nowadays available to more and more people.

This makes it all the more important to handle and process these data carefully. For this purpose, statistics is just as indispensable today as it has always been. Of course, the current explosion in the quantity of data also brings along previously unknown problems regarding data management. In this respect, statisticians rely on the support of IT-experts. When it comes to data analysis, however, the basic principles of estimating and testing, model selection, or sampling apply to big data sets (e.g. supplied by Google) just as they do to smaller data sets (e.g. from the Federal Statistical Office).

This volume presents selected studies from the Statistics department at the TU Dortmund University, the only independent statistics department in Germany, as well as from former DFG (German Research Foundation) Collaborative Research Centers associated with the department. In addition, a number of researchers from outside the TU have contributed valuable research shedding light on how statistics can help us unveil and understand phenomena of our daily lives.

This book shows, for example, how statistics can help to assess the effect of drugs; how statistics can be utilized to analyze musical audio data to automate transcriptions; how flood catastrophes or risks on the stock market can be statistically modeled and thus better managed; and how meaningful quantitative data can be extracted from qualitative information such as texts and spoken words. Other studies in this volume deal with quality control in industry, the forecasting abilities of rating agencies, or the prevention of false alarms in intensive care. Additionally, the volume elucidates why the lottery is not purely a game of chance and why favorites are systematically overrated in horse betting.

We are reporting cutting edge research accessible to non-statisticians and non-experts in the related fields. The editors asked their participating colleagues to avoid using specialist language as much as possible. The volume is geared towards an audience that wants to gain insights into general ideas about statistical applications but does not strive to understand all the subtleties of formal statistical analysis. We hope that this book will therefore also appeal to those whose enthusiasm for statistics has somewhat suffered in their degree courses in business administration, economics, psychology, sociology, or other subjects where statistics certificates are traditionally required. Unfortunately, in academic education the intrinsic beauty of statistics is often obscured by too much formalism. This anthology will hopefully demystify the specter of statistics and invite you to an unobstructed view of an utterly fascinating discipline.

Preparation of the English Version

In most parts, this book is a translation of the Springer book "Faszination Statistik", which was originally published in German. Only Chap. 25 ("Statistical modeling of current linguistic realities around the world: The case of Singapore") and Chap. 26 ("Linguistic manifestations of cultural differences across national varieties of English – a methodological survey") were explicitly written for the English version. A first translation of the chapters in "Faszination Statistik" was produced by the corresponding authors, partly supported by the algorithms DeepL (https://www.deepl.com/translator) or Google Translate (https://translate.google.com/). Those translations were reviewed concerning linguistic expression and clarity of content by Sarah Buschfeld, one of the book editors. She also suggested modifications and additions to the corresponding authors whenever necessary. Claus Weihs, first editor of this volume, discussed the suggested changes with the corresponding authors, checked the comprehensibility of statistical contents, and partly reformatted the chapters to unify their appearance. Finally, the texts were approved by the corresponding authors.

Acknowledgment

The editors would like to thank Brian Hess for his support in proofreading some of the chapters.

Dortmund, Germany   Claus Weihs
September 2023   Walter Krämer
                                                                      Sarah Buschfeld

Contents

Part I   Human Life, Medicine, and Genetics

1   Season of Birth and Human Longevity: A New Theory Why Children Born in November Live Longer 3
Walter Krämer and Katharina Schüller
1.1   The Date of Birth Matters 3
1.2   A New Look from Switzerland 5
1.3   In Search of the Reasons 7
1.4   Data Limitations 10
1.5   Further Reading 10

2   Where Do Drugs Work in the Body? A Systematic Statistical Data Analysis 11
Claus Weihs
2.1   Pharmacokinetics and the Pre-clinical Phase 11
2.2   Standard Procedure for Statistical Data Analysis 12
2.3   The Distribution of Drugs in the Body 13
        2.3.1   Business Understanding 13
        2.3.2   Data Understanding 13
        2.3.3   Data Preparation 15
        2.3.4   Modeling 16
        2.3.5   Evaluation 18
        2.3.6   Deployment 18
2.4   Further Reading 19

3   Drug Studies: Using Statistics to Achieve the Optimal Dose 21
Holger Dette and Kirsten Schorning
3.1   The Three Clinical Test Phases 21
3.2   The Optimization of Phase 2 22
3.3   It is All About the Experimental Design 23
3.4   Towards Practical Application 26
3.5   Further Reading 26

4   Statistical Alarm Systems in Intensive Care Medicine   27
      Roland Fried, Ursula Gather, and Michael Imhoff
      4.1   Alarms in Acute Medical Care   27
      4.2   Smoothing as Part of Data Preprocessing   29
      4.3   Joint Analysis of Characteristics   31
      4.4   Validation of the Results   34
      4.5   Further Reading   34

**5   Personalized Medicine: How Statistics Helps Not to Drown
      in the Flood of Data**   37
      Jörg Rahnenführer
      5.1   Genetic Decision Support in Medicine.............................   37
      5.2   Efficacy and Side Effects of Therapies   38
      5.3   In Search for Genetic Patterns   39
      5.4   Statistical Tricks...   40
      5.5   Medical Application...   41
      5.6   Models for Estimating Disease Progression   41
      5.7   Summary..   43
      5.8   Further Reading   43

**6   Modulating Genetic Effects on Bladder Cancer Risk in an
      Area of Coal, Iron, and Steel Industries**   45
      Silvia Selinski, Katja Ickstadt, and Klaus Golka
      6.1   Environment, Genes, and Urinary Bladder Cancer   45
      6.2   Epidemiology and Genetics..   47
      6.3   Gene-Environment Interactions   50
      6.4   Conclusion...   52
      6.5   Further Reading   53

7   Statistics and the Maximum Human Lifespan   55
      Jan Feifel and Markus Pauly
      7.1   Background...   55
      7.2   From the Average to Extreme Value Theory   56
      7.3   Challenges in Working with Demographic Data....................   59
      7.4   Results   60
      7.5   Conclusion...   60
      7.6   Further Reading   61

Part II   Sports and Entertainment

8   Statistics and Soccer..   65
      Andreas Groll and Gunther Schauberger
      8.1   More Goals by Means of Statistics   65
      8.2   A Statistical Model for Predicting Goals...........................   66
      8.3   Influential Variables   67
      8.4   Conclusion...   72
      8.5   Further Reading   72

**9   The Players' Anxiety at the Penalty Kick: Who Is the Best
      Penalty Taker, Who the Best Goalkeeper?** 73
      Peter Gnändinger, Leo N. Geppert, and Katja Ickstadt
      9.1   Penalties in Soccer 73
      9.2   The Penalty Data Set 74
      9.3   Factors Associated with the Outcome of the Penalty 75
             9.3.1   Modeling Penalty Probabilities 75
             9.3.2   Fixed and Random Effects 77
             9.3.3   Important Factors 78
      9.4   Rankings of Goalkeepers and Penalty Takers 79
             9.4.1   Penalty Takers' Influence as Random Effect 79
             9.4.2   Leaderboards 80
      9.5   Conclusion and Outlook 82
      9.6   Further Reading 82

10   Music Data Analysis 83
      Claus Weihs
      10.1   What Is Music? 83
      10.2   Music Data 85
      10.3   The Studies 87
             10.3.1   Classification 87
             10.3.2   Pitch Identification 87
             10.3.3   Instrument Recognition 89
             10.3.4   Onset Detection 89
             10.3.5   Automatic Transcription 90
             10.3.6   Genres 91
      10.4   Further Reading 92

11   Statistics and Horse Race Betting: Favorites vs. Longshots 93
      Martin Kukuk
      11.1   Horse Race Betting 93
      11.2   Betting Payouts 94
      11.3   Empirical Explanations for the Favorite-Longshot Bias 95
      11.4   Favorite-Longshot-Bias Caused by Subjective Estimates 96
      11.5   Conclusion 98
      11.6   Further Reading 99

12   The Statistics of the German 6/49 Lotto 101
      Walter Krämer
      12.1   Lotto as an Investment 101
      12.2   Optimizing the Payout 103
      12.3   Further Reading 105

Part III   Money and Business

13   Statistics at the Stock Exchange 109
Walter Krämer and Tileman Conring
   13.1   Beware of Dependencies 109
   13.2   Investing in Stocks 110
   13.3   Time-Varying Dependencies 111
   13.4   The Not So Normal Normal Distribution 113
   13.5   Cointegration 115
   13.6   Further Reading 116

14   Statistics in the Risk Assessment of Bank Portfolios 117
Dominik Wied and Robert Löser
   14.1   The Problem 117
   14.2   Expected Shortfall Compared to Value-at-Risk 118
   14.3   Estimation of Risk Measures 119
   14.4   Validation of Risk Models 120
   14.5   Further Reading 123

15   On Rating the Raters: Statistics in the Rating Industry 125
Walter Krämer and Simon Neumärker
   15.1   Obligations and Obligors 125
   15.2   How to Judge the Quality of Default Forecasts? 127
   15.3   A Numerical Example 129
   15.4   Partial Orderings of Probability Forecasts 130
   15.5   Scalar Valued Measures of Forecasting Quality 130
   15.6   Further Reading 132

**16   Gross Domestic Product, Greenhouse Gas Emissions, and
     Global Warming** 133
Martin Wagner, Fabian Knorre, and Christina Kopetzky
   16.1   Economic Activity and Emissions 133
   16.2   Statistical Analysis 136
   16.3   Parameter Estimation in the Presence of Nonlinear
        Cointegration 138
   16.4   Interpretation 140
   16.5   Further Reading 140

Part IV   Nature and Technology

17   Flood Statistics: Still on the River Bank or Already in the Water? ... 143
Svenja Fischer, Roland Fried, and Andreas Schumann
   17.1   Getting a Grip on Floods 143
   17.2   What Is a Flood? 145
   17.3   Flood Risk and Probabilities 146
   17.4   Robust Estimation 148

17.5 Flood Types and Changes Over Time 149
17.6 Regionalization 151
17.7 Further Reading 152

18 **How Statistics Helps to Reduce Rejects** 153
Claus Weihs and Nadja Bauer
18.1 Defects in Deep Drilling 153
18.2 Quality Improvement: Six Sigma................................... 154
18.2.1 Define 155
18.2.2 Measure 156
18.2.3 Analyze 158
18.2.4 Improve 160
18.2.5 Control 160
18.3 Further Reading 161

19 **Statistics and Reliability of Technical Products** 163
Christine H. Müller
19.1 Reliability and Randomness 163
19.2 Simple Service Lifetime Analysis 164
19.3 Lifetime Analysis Under Different Loads........................... 165
19.4 Lifetime Analysis for Products with Several Components 166
19.5 Prediction Intervals... 167
19.6 Outlook 169
19.7 Further Reading 169

20 **Durable Machine Components: How Statistical Design of
Experiments Optimizes Wear Protection** 171
Sonja Kuhnt, Wolfgang Tillmann, Alexander Brinkhoff,
and Eva-Christina Becker-Emden
20.1 Wear Protection Through Coating 171
20.2 Optimization Through Statistical Design of Experiments.......... 172
20.3 Challenges in Real Coating Processes.............................. 175
20.4 Further Reading 177

Part V Intricacies of Measurement

21 **Measuring the Immeasurable: Statistics, Intelligence,
and Education**... 181
Philipp Doebler, Gesa Brunn, and Fritjof Freise
21.1 Educational Tests and Education 181
21.2 Latent Variables and Their Indicators 182
21.3 A Statistical Model for Learning Progress Diagnostics 185
21.4 From Data to Latent Variables 186
21.5 Further Reading 189

22   Uncovering Embarrassing Truths Through Statistics 191
     Andreas Quatember
     22.1   The Method of Indirect Questioning 191
     22.2   A Modification of the Original Idea 192
     22.3   Tasks for Future Research 194
     22.4   Further Reading 195

23   Samples and Missing Data 197
     Andreas Quatember
     23.1   Sampling in Theory and Practice 197
     23.2   Statistical Methods to Compensate for Non-responses 198
     23.3   Further Reading 200

Part VI   Language Data

**24   Who Is Supposed to Read All This? Automatic Analysis of
     Text Data** 203
     Jörg Rahnenführer and Carsten Jentsch
     24.1   Large Text Collections 203
     24.2   Text Analysis in the Social Sciences 204
     24.3   Preprocessing of Text Data 205
     24.4   Topic-Based Classification of Large Text Collections 206
     24.5   Finding Differences 207
     24.6   Text Analysis of Election Programs 208
     24.7   Summary and Outlook 211
     24.8   Further Reading 211

**25   Statistical Modeling of Current Linguistic Realities Around
     the World: The Case of Singapore** 213
     Sarah Buschfeld and Claus Weihs
     25.1   Singapore and the World Englishes Paradigm 213
     25.2   Data Collection and Preparation 214
     25.3   Prediction of Linguistic Characteristics 216
             25.3.1   Design of Experiments 216
             25.3.2   Variation in British and Singaporean Englishes 217
     25.4   Evaluation and Interpretation 221
     25.5   Summary and Outlook 222
     25.6   Further Reading 222

**26   Linguistic Manifestations of Cultural Differences Across
     National Varieties of English: A Methodological Survey** 225
     Edgar W. Schneider
     26.1   Introduction 225
     26.2   Electronic Corpora: Representative Text Collections? 227
     26.3   Linguistic Forms Representing Cultural Orientations 230
             26.3.1   Example of a Cultural Dimension: Collectivism 230

26.3.2 Example of a Structural Scheme: Recipientless
                        Constructions 232
        26.4   Statistical Testing... 233
        26.5   Sample Results 234
        26.6   Summary and Conclusion.. 235
        26.7   Further Reading 235

Part VII   From Here to Where?

27   Is Data Science More Than Statistics? The Bigger Picture 239
        Claus Weihs and Katja Ickstadt
        27.1   Data Science: What Is It Anyway? 239
        27.2   Data Science: Steps 240
                27.2.1 General Structure 240
                27.2.2 Data Acquisition and Enrichment......................... 241
                27.2.3 Data Exploration... 243
                27.2.4 Modeling: Statistical Data Analysis 244
                27.2.5 Evaluation: Model Validation and Selection 245
                27.2.6 Deployment of Results 246
        27.3   Conclusion... 246
        27.4   Further Reading 246

Index.. 249

Contributors

Nadja Bauer [Chapter 18] FH Dortmund, Faculty of Computer Science, Dortmund, Germany nadja.bauer@fh-dortmund.de

Eva-Christina Becker-Emden [Chapter 20] FH Dortmund, Faculty of Computer Science, Dortmund, Germany eva-christina.becker-emden@fh-dortmund.de

Alexander Brinkhoff [Chapter 20] TU Dortmund, Dortmund, Germany alexander.brinkhoff@tu-dortmund.de

Gesa Brunn [Chapter 21] Deutscher Wetterdienst, Research and Development, Offenbach, Germany gesa-marie.brunn@dwd.de

Sarah Buschfeld [editor, Chapter 25] TU Dortmund, Department of Cultural Studies, Dortmund, Germany sarah.buschfeld@tu-dortmund.de

Tileman Conring [Chapter 13] TU Dortmund, Department of Statistics, Dortmund, Germany tileman.conring@tu-dortmund.de

Holger Dette [Chapter 3] RU Bochum, Stochastics, Bochum, Germany holger.dette@ruhr-uni-bochum.de

Philipp Doebler [Chapter 21] TU Dortmund, Department of Statistics, Dortmund, Germany doebler@statistik.tu-dortmund.de

Jan Feifel [Chapter 7] Ulm, Germany jan.feifel@alumni.uni-ulm.de

Svenja Fischer [Chapter 17] RU Bochum, Engineering Hydrology and Water Resources Management, Bochum, Germany svenja.fischer@ruhr-uni-bochum.de

Fritjof Freise [Chapter 21] University of Veterinary Medicine Hannover, Department for Biometry, Epidemiology and Information Processing, Hannover, Germany fritjof.freise@tiho-hannover.de

Roland Fried [Chapters 4, 17] TU Dortmund, Department of Statistics, Dortmund, Germany msnat@statistik.tu-dortmund.de

Ursula Gather [Chapter 4] TU Dortmund, Department of Statistics, Dortmund, Germany ursula.gather@tu-dortmund.de

Leo N. Geppert [Chapter 9] TU Dortmund, Department of Statistics, Dortmund, Germany geppert@statistik.tu-dortmund.de

Peter Gnändinger [Chapter 9] TU Dortmund, Department of Statistics, Dortmund, Germany peter.gnaendinger@tu-dortmund.de

Klaus Golka [Chapter 6] IfADo, Clinical Occupational Medicine, Dortmund, Germany golka@ifado.de

Andreas Groll [Chapter 8] TU Dortmund, Department of Statistics, Dortmund, Germany groll@statistik.tu-dortmund.de

Katja Ickstadt [Chapters 6, 9, 27] TU Dortmund, Department of Statistics, Dortmund, Germany ickstadt@statistik.tu-dortmund.de

Michael Imhoff [Chapter 4] RU Bochum, Faculty of Medicine, Bochum, Germany mike@imhoff.de

Carsten Jentsch [Chapter 24] TU Dortmund, Department of Statistics, Dortmund, Germany jentsch@statistik.tu-dortmund.de

Fabian Knorre [Chapter 16] TU Dortmund, Department of Statistics, Dortmund, Germany knorre@statistik.tu-dortmund.de

Christina Kopetzky [Chapter 16] University of Klagenfurt, Department of Economics, Klagenfurt, Austria christina.kopetzki@aau.at

Walter Krämer [editor, Chapters 1, 12, 13, 15] TU Dortmund, Department of Statistics, Dortmund, Germany walterk@statistik.tu-dortmund.de

Sonja Kuhnt [Chapter 20] FH Dortmund, Faculty of Computer Science, Dortmund, Germany sonja.kuhnt@fh-dortmund.de

Martin Kukuk [Chapter 11] University of Würzburg, Department of Economics, Würzburg, Germany martin.kukuk@uni-wuerzburg.de

Robert Löser [Chapter 14] TU Dortmund, Dortmund, Germany robert.loeser@tu-dortmund.de

Christine H. Müller [Chapter 19] TU Dortmund, Department of Statistics, Dortmund, Germany cmueller@statistik.tu-dortmund.de

Simon Neumärker [Chapter 15] TU Dortmund, Department of Statistics, Dortmund, Germany simon.neumaerker@tu-dortmund.de

Markus Pauly [Chapter 7] TU Dortmund, Department of Statistics, Dortmund, Germany markus.pauly@tu-dortmund.de

Andreas Quatember [Chapters 22, 23] JKU Linz, Institute of Applied Statistics, Linz, Austria andreas.quatember@jku.at

Jörg Rahnenführer [Chapters 5, 24] TU Dortmund, Department of Statistics, Dortmund, Germany rahnenfuehrer@statistik.tu-dortmund.de

Gunther Schauberger [Chapter 8] TU Munich, Epidemiology, Munich, Germany gunther.schauberger@tum.de

Edgar W. Schneider [Chapter 26] University of Regensburg, English Linguistics, Regensburg, Germany edgar.schneider@sprachlit.uni-regensburg.de

Kirsten Schorning [Chapter 3] TU Dortmund, Department of Statistics, Dortmund, Germany schorning@statistik.tu-dortmund.de

Katharina Schüller [Chapter 1] Stat-Up Statistical Consulting & Data Science GmbH, Munich, Germany katharina.schueller@stat-up.com

Andreas Schumann [Chapter 17] RU Bochum, Engineering Hydrology and Water Resources Management, Bochum, Germany andreas.schumann@hydrology.ruhr-uni-bochum.de

Silvia Selinski [Chapter 6] IfADo, Toxicology/Systemtoxicology, Dortmund, Germany selinski@ifado.de

Wolfgang Tillmann [Chapter 20] TU Dortmund, Department of Mechanical Engineering, Dortmund, Germany wolfgang.tillmann@udo.edu

Martin Wagner [Chapter 16] University of Klagenfurt, Department of Economics, Klagenfurt, Austria

Bank of Slovenia, Ljubljana, Slovenia martin.wagner@aau.at

Claus Weihs [editor, Chapters 2, 10, 18, 25, 27] TU Dortmund, Department of Statistics, Dortmund, Germany claus.weihs@tu-dortmund.de

Dominik Wied [Chapter 14] University of Cologne, Institute for Statistics and Econometrics, Köln, Germany dwied@uni-koeln.de

Part I
Human Life, Medicine, and Genetics

Chapter 1
Season of Birth and Human Longevity: A New Theory Why Children Born in November Live Longer

Walter Krämer and Katharina Schüller

Abstract In Europe, human individuals born in November live on average half a year longer than individuals born in May. The ultimate cause must involve the climate, because in the southern hemisphere, this pattern is reversed. However, it is still unclear how exactly environmental factors related to the season of birth affect longevity. We offer an explanation.

1.1 The Date of Birth Matters

Since ancient times, astrologers have tried to deduce the further fate of a person from the date of birth. The fact that such influences exist helps them a lot. However, these influences have nothing to do with planets or with stars. It has for instance been known for decades that individuals born in February or March suffer an increased risk of schizophrenia. Or consider professional sports. Psychologist Peter Jensen once recorded the month of birth of Canadian national hockey players. They were all born in January, February, March, or April. None of them was born in May, June, July, August, September, October, November, or December. In Australia, it has likewise been found that in almost all professional sports there are more successful athletes born in January than born in December. However, this does not result from the stars but from the fact that, in cohorts grouped by year of birth as is usual in many sports, children born in January are the oldest and for this reason often the best in their respective training groups. Therefore, they also receive a disproportionate amount of attention and support. If this happens many years in a row until the option

W. Krämer (✉)
TU Dortmund, Department of Statistics, Dortmund, Germany
e-mail: walterk@statistik.tu-dortmund.de

K. Schüller
Stat-Up Statistical Consulting & Data Science GmbH, Munich, Germany
e-mail: katharina.schueller@stat-up.com

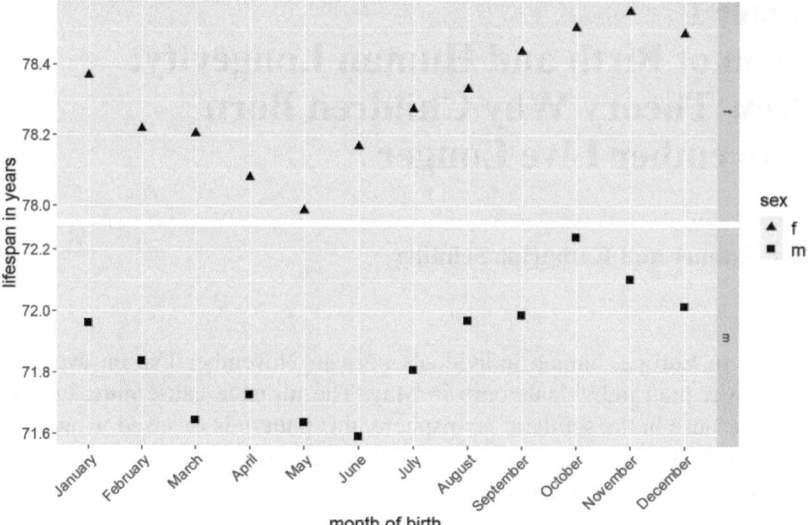

Fig. 1.1 Average length of life of all Swiss individuals who died between 1969 and 2010. Lower panel males, upper panel females

of a professional career in sports is entertained, it is no wonder that children born in January are privileged when entering the professional career.

Here, the date of birth affects success in the job. But the date of birth also affects longevity. This has been known for decades and has been documented for many countries. Figure 1.1 illustrates this for Switzerland. It shows the average age at death of all Swiss individuals who died between 1969 and 2010, grouped by month of birth. As can be seen, for men and women alike, individuals born in October/November live on average six months longer than those born in May.

Figure 1.1 does not provide a realistic picture of life expectancy in Switzerland. It is an underestimation, because the final years of our data set include a disproportionate amount of individuals who died rather young. The seasonal effect in longevity however is obvious. And it is not produced by chance. Similar results have been shown for Denmark, Sweden, or Germany, albeit based on smaller data sets. On the southern hemisphere, the opposite occurs. Here, individuals born in the fall suffer a loss in life expectancy. Therefore, it is obvious that temperature or sun exposure during conception or in-utero or early postnatal periods must somehow be involved. For example, longevity of Australians who have immigrated from Europe follows the pattern of the northern hemisphere. However, it is still not clear how exactly season of birth connects to length of life. Here we offer an explanation.

1.2 A New Look from Switzerland

Our analysis is based on a unique data set of all recorded cases of death in Switzerland from 1969 to 2010, including date, place and cause of death, kindly provided by the Swiss Federal Statistical Office. The data also include date and place of birth plus information about spouse, parents, religion, language region, profession, and economic status. Figure 1.2 shows the average length of life of individuals who have died of cancer from 1969 to 2010, and Fig. 1.3 shows the average length of life of individuals who, in the same time span, have died from cardiovascular diseases. For both causes of death, and for men and women alike, the seasonal pattern of Fig. 1.1 repeats itself.

Figure 1.4, on the other hand, shows the average length of life of individuals who have died in an accident. For obvious reasons, their lives were shorter, and, most importantly, the seasonal pattern disappears. This is what one would expect, as seasonality in longevity must somehow be encoded in the human body and will therefore be annihilated by any independent random cause of death. It might also be of interest that the average age at death of male victims of accidents is about 15 years below that of females.

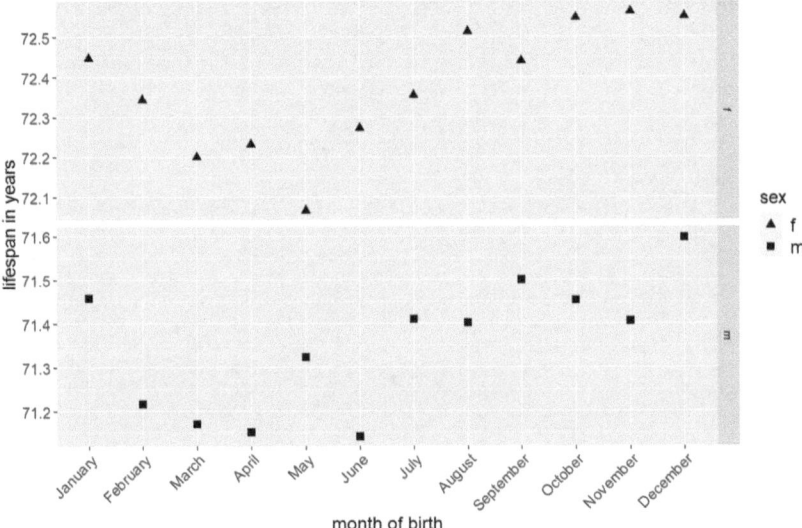

Fig. 1.2 Average length of life of all Swiss individuals who died from cancer between 1969 and 2010. Upper panel females, lower panel males

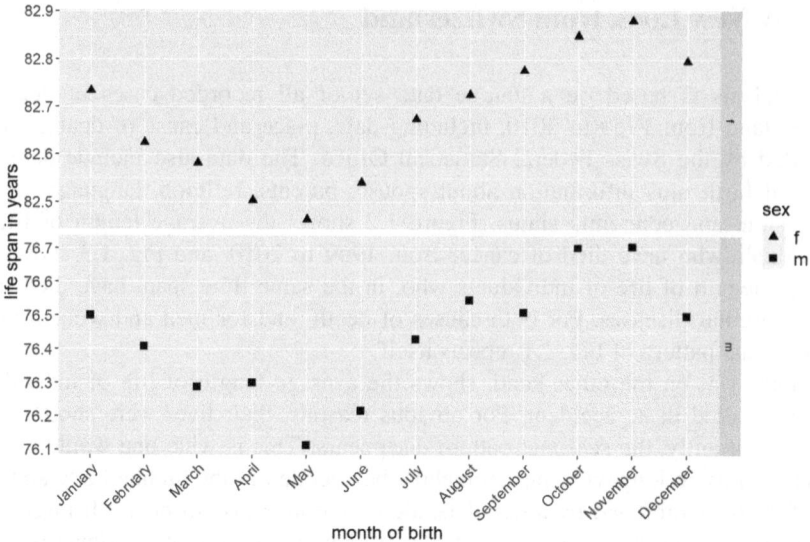

Fig. 1.3 Average length of life of all Swiss individuals who died from cardiovascular diseases between 1969 and 2010. Upper panel females, lower panel males

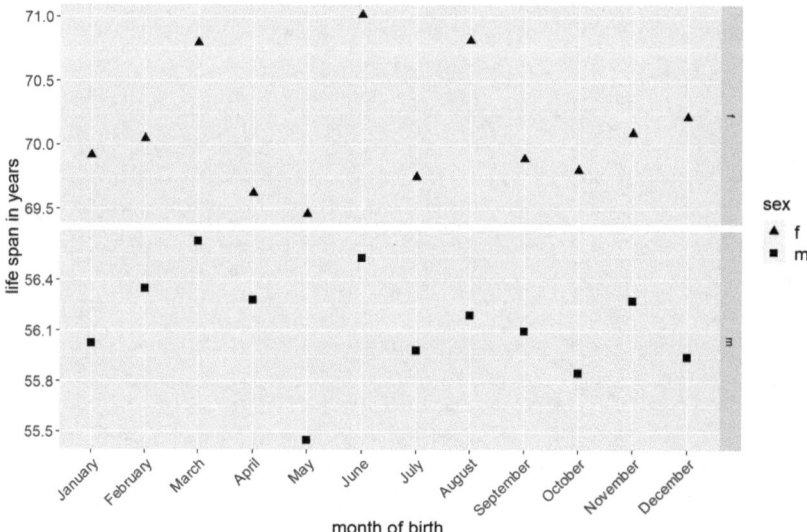

Fig. 1.4 Average length of life of all Swiss individuals who died from accidents between 1969 and 2010. Upper panel females, lower panel males

1.3 In Search of the Reasons

For a long time, solar radiation has been a prime suspect for influencing length of life. Consider the length of life of all members of the US Congress born between 1750 and 1900. This does not increase monotonously over time, as one would expect from the cumulative character of advances in hygiene and medicine, but fluctuates in cycles of nine to twelve years. For example, the average length of life of members of congress born in and after 1752 rises up to age cohort 1763 and then starts to decrease. These cycles correspond almost exactly to those of sunspots and thus to the intensity of the sun's radiation. However, the exact mechanism behind this effect is still unknown.

One possible cause is different levels of vitamin D in the baby, caused by different exposures to sunlight of the mother. This might also explain the seasonal effect which is the focus of interest here: If, in the critical final months prior to birth, mothers are exposed to sunlight, this will increase production of vitamin D and thus the health of babies. Other studies point out that parents often have more stress in winter and are more prone to infectious diseases, with lasting effects on babies born soon after, while in summer, mothers are more relaxed, taking time off for holidays and make life more comfortable for the unborn, again with lasting effects for their later life.

Here we explore an explanation brought into the spotlight by a recent article in Nature Medicine that not the mother, but the father might be the driving force behind the seasonal effects in the length of life of the child. And indirectly, via the ambient temperature at the time of conception, the sun is involved again. However, the effect is neither caused by vitamins nor fresh air nor mental stress but by a very special kind of body substance called brown adipose tissue. This rather unpleasant name refers to a special and extremely useful type of body tissue whose cells are able to generate heat through oxidation. It is common in newborns, who lose a lot of heat due to a larger body surface when compared to body volume, and is found primarily on the neck and on the chest. It was believed that this tissue disappears with age. However, it was discovered recently that this extremely health-promoting regulator also occurs in the body of adults, albeit in smaller quantities, and that its occurrence negatively correlates with ambient temperature at the time of conception—the colder, the higher the amount of brown adipose tissue in the body later in life. It is as if the sperms of the father were transmitting a message to the embryonic cells: "Produce lots of brown adipose tissue, this is good for you."

Could this be the reason for the longer life of children born in November? They are conceived in the coldest months of the year, in January and February and therefore spend their lives with more brown adipose tissue than children conceived in the summer.

Below we further explore this hypothesis. Our data is from Switzerland, a transition zone between three climate zones: the Atlantic maritime climate, the continental climate and the Mediterranean climate; in the high mountains there are even polar conditions. The Mediterranean climate dominates south of the Alps,

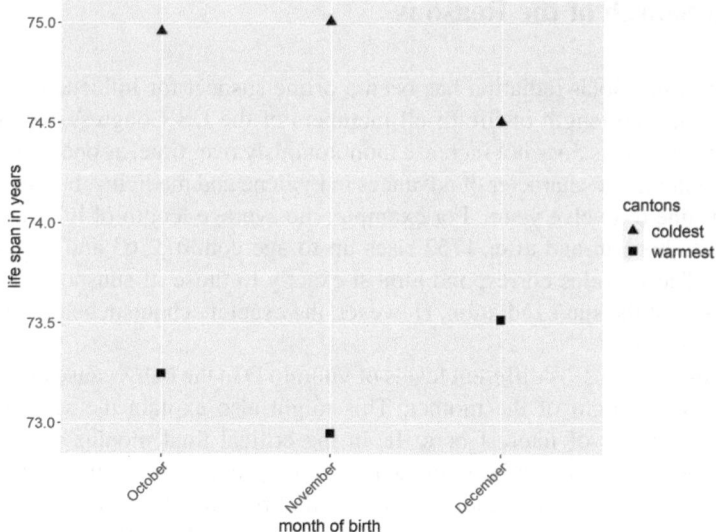

Fig. 1.5 Average length of life for individuals born in warm and cold cantons. Upper panel: cold; lower panel: warm

with much milder winters. Taken together, this implies a wide range of ambient temperatures at birth. As we only know the date of birth, not the date of conception, we assume an average duration for all pregnancies. We also assume that cantons of birth and conception are identical. While this is not correct in any case, it will still hold for a large majority.

Figure 1.5 shows the average length of life for individuals born in the three coldest and in the three warmest cantons of Switzerland for the months of October, November, and December: People obviously live longer when it is cold outside during conception.

In another study we have grouped all cases of death according to average ambient temperature in the month of conception in the canton of birth. This approach no longer distinguishes whether a child was conceived in January or July, but only whether there were positive or negative temperatures in the month of conception and how extreme they were (assuming that canton of birth and canton of conception are identical). It is seen that the difference in mean longevity is now even larger compared to the grouping according to month of birth, lending further strength to our hypothesis (cf. Fig. 1.6).

Figure 1.7 repeats this study for all individuals born before 1960. This is to eliminate a potential effect of global warming, which, in our data set, produces the effect that among individuals born in later and warmer years, only those are included in our data set who died rather young. But, as the figure shows, the effect persists. The major difference is that average length of life is now larger as early deaths have been eliminated.

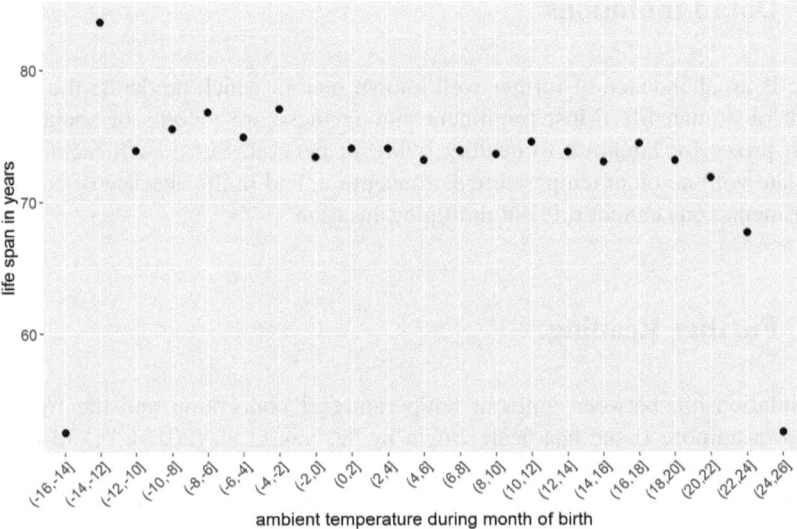

Fig. 1.6 Average length of life of Swiss individuals conceived in different ambient temperatures

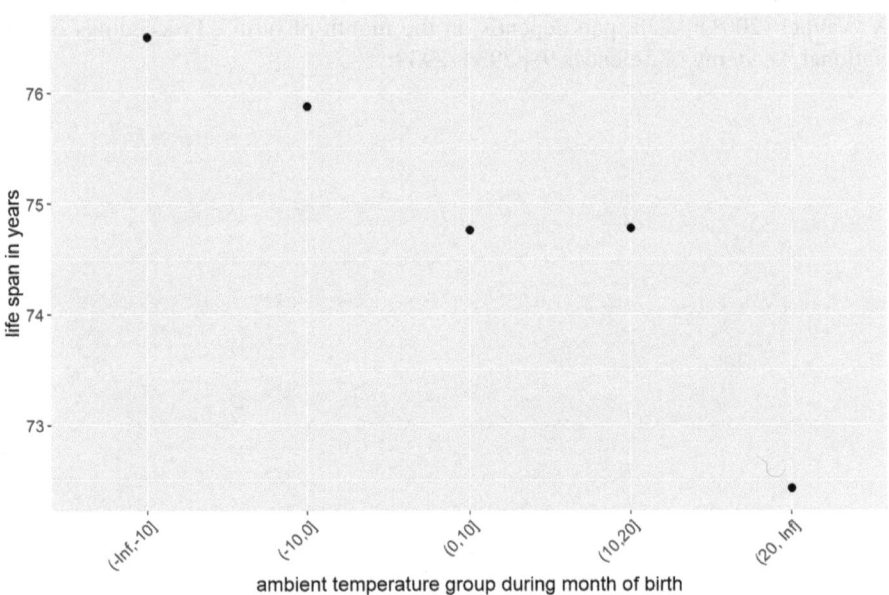

Fig. 1.7 Average length of life of Swiss individuals born before 1960 which are conceived in different ambient temperatures

1.4   Data Limitations

There is an abundance of further well known factors which moderate the average length of human life. Most prominent among these are income or social status, which proxy for attention to healthy living in general. Since such factors might correlate with ambient temperature at conception, and in the absence of controlled experiments, one cannot rule out multiple causation.

1.5   Further Reading

The relationship between ambient temperature at conception and the frequency of brown adipose tissue has been shown by W. Sun et al. (2018): "Cold-induced epigenetic programming of the sperm enhances brown adipose tissue activity in the offspring," Nature Medicine 24, 1372–1383. On the relationship between solar cycles and life expectancy, see David A. Juckett and Barnett Rosenberg (1993): "Correlation of Human Longevity Oscillations with Sunspot Cycles," Radiation Research 133(3), 312–320. The best analysis of the relationship between month of birth and longevity to date is provided by Gabriele Doblhammer and James W. Vaupel (2001): "Lifespan depends on the month of birth", Proceedings of the National Academy of Sciences 98, 2934–2939.

Chapter 2
Where Do Drugs Work in the Body?
A Systematic Statistical Data Analysis

Claus Weihs

Abstract The distribution of drugs in the human body is one of the decisive factors for the success of a therapy. With the help of statistical analyses and by means of an example, we show that, for important therapeutic goals, drugs actually reach the desired organs.

2.1 Pharmacokinetics and the Pre-clinical Phase

'Pharmacokinetics' is concerned with the entirety of the processes in which medical substances are disseminated in the body, from their ingestion, to distribution, biochemical conversion, and degeneration to excretion. Here, we are particularly interested in the distribution of medical substances in the body. This process starts immediately after ingestion and is influenced by, among other things, the solubility of the substance, its chemical structure, and its ability to bind to proteins. Fat-soluble substances, for example, tend to accumulate in adipose tissue. In addition, the organ or tissue perfusion, the pH-value in the tissue or body fluid, and the permeability of the membranes involved play important roles.

Before a drug is tested on humans, it undergoes various pre-clinical tests. A potentially active agent is tested on bacteria, cell and tissue cultures, in animal experiments, or on isolated organs. Distribution tests on drug candidates are an essential part of such testing. In general, pre-clinical tests take two years. Drug candidates that do not meet the requirements are discarded. Only compounds that successfully pass all tests are released for human subject research.

We are interested in how the distribution of drugs in the body differs for different therapeutic goals. This is an ideal case for a systematic statistical data analysis.

C. Weihs (✉)
TU Dortmund, Department of Statistics, Dortmund, Germany
e-mail: claus.weihs@tu-dortmund.de

C. Weihs et al. (eds.), *Statistics Today*, Society, Environment and Statistics,
https://doi.org/10.1007/978-3-662-68907-3_2

11

2.2   Standard Procedure for Statistical Data Analysis

The de facto standard procedure for systematic applied statistical data analysis is
CRISP-DM (Cross Industry Standard Process for Data Mining), which is organized
in six main steps:

Business understanding, data understanding, data preparation, modeling,
evaluation, and deployment (see Fig. 2.1).

CRISP-DM was developed since 1996 by the companies NCR (National Cash
Register), SPSS (Statistical Package for the Social Sciences), and DaimlerChrysler
and is widely used today. The companies pursued this development for different reasons: NCR was mainly interested in the added value for data warehouse customers,
SPSS in a concept for the data mining product Clementine, and DaimlerChrysler in
the combination of practical experience and conceptual considerations.

CRISP-DM is not a theoretical academic product based on technical principles.
It was not invented behind closed doors. Instead, it originates from practical
experience with real-life problems. In fact, applied statisticians today (almost)
automatically follow the steps of the CRISP-DM process.

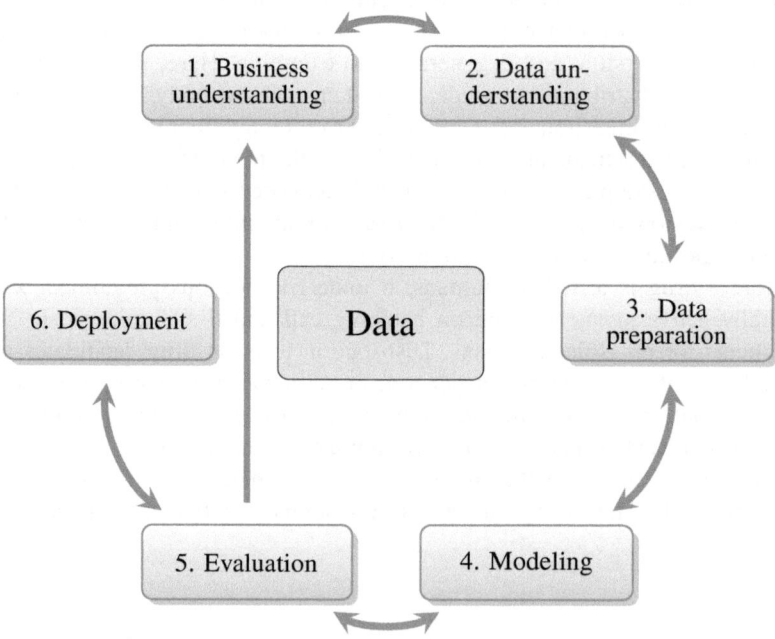

Fig. 2.1 CRISP-DM process

2.3 The Distribution of Drugs in the Body

In the following, we apply CRISP-DM to the analysis of data on the distribution of drugs in the human body. For this purpose, we proceed as illustrated in Fig. 2.1.

2.3.1 Business Understanding

The first step of CRISP-DM is to understand the problem. To that end, we translate the content-related objectives into objectives that appear to be practically realizable from the data analysis perspective. We focus on the prediction of therapy classes, such as antihypertensive or antidepressant/neuroleptic, according to the distribution pattern in the body, i.e. on the question

How well can we predict therapy classes from the distribution of the drug in the body?

Thus, the goal of our data analysis is to find a relationship between distribution patterns and therapy classes. To reach this goal, we first decide how to deal with missing values, outliers, and domain knowledge. Then, we use a so-called decision tree to determine the relationship between the therapy classes and the concentration of the drug in the organs. Classification by decision trees allows for a fairly simple interpretation of class assignments (see Sect. 2.3.4).

2.3.2 Data Understanding

The next step of CRISP-DM is to understand the data, i.e. the data acquisition process and the raw data, including the identification of data-related problems and gaining first insights into the data structure.

In our example, the data consist of measurements of radioactive-labeled substances in the different organs of rats five to six minutes after intravenous injection. This particular time of investigation ensures that the distribution of the substance is complete and that metabolization, i.e. the biochemical conversion and degeneration of a drug in the body, plays a subordinate role. By following this procedure, we hope to unveil a direct relationship between therapy class and distribution pattern.

The desired therapy class of the injected drugs is known from domain knowledge. Raw data are available for 20 individual tests S1, ..., S20 including different substances from different therapy classes (T) (see Table 2.1). The analysis is normally performed on three to four (male) rats per test and 85 rats in total.

Table 2.1 Assignment of the individual tests S1, ..., S20 to therapy classes

Class T	Type of therapy	Tests
1	Anti-inflammatory	S1, S3, S7, S9
2	Antidepressants and neuroleptics	S4, S5, S10, S11, S18
3	Beta blockers and ca-antagonists	S6, S8, S16
4	Cytostatic agents/antitumor	S2, S14, S15, S20
5	Biopolymers	S12, S13, S17, S19
7	All classes except 2 and 3	as in T = 1, 4, 5

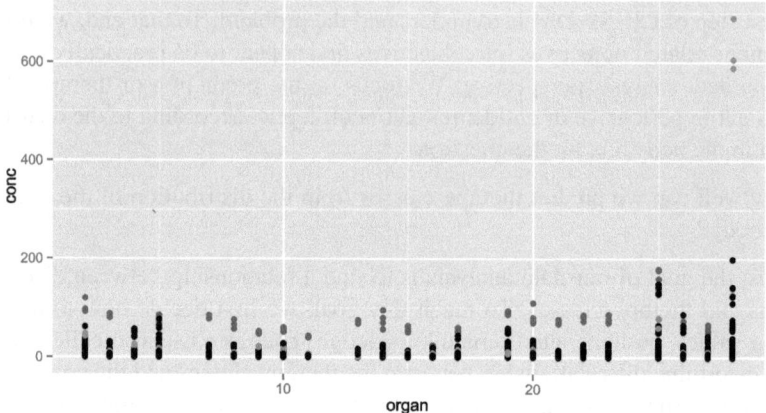

Fig. 2.2 Concentration of drugs (conc) (in total radioactivity) according to organ label (organ). The organs are numbered consecutively from 1 = blood to 24 = kidney; the spaces in the visualization after every fifth organ are intended to increase the comprehensibility of the results. For example, organ 24 is numbered 28 on the x-axis. The results from test S19 are presented in blue, the results from S2 in orange

Radioactive concentrations are measured in at least 24 organs, partitioned into four groups of five organs each and one group with four organs:

Blood circulation: blood, plasma, heart, aorta, lung
Other organs: spleen, muscles, skin, sciatic nerve, eye
Adipose tissue: brain, bone marrow, testicles, white and brown fat
Glands: adrenal glands, thyroid, thymus, salivary gland, pancreas
Digestive organs: liver, stomach, small intestine, kidney.

Below, the organs are labeled according to their order in this list, e.g. blood is organ 1 (O1), heart organ 3 (O3), and brain organ 11 (O11). The aim is to determine the relationship between the therapy class and the concentration of the drugs in these organs.

Figure 2.2 presents the results of a first descriptive analysis of the concentration of drugs in the different organs. The highest concentrations can be found for kidney in test S19. In test S2, the concentrations are particularly high overall. This leads to the hypothesis that the overall level of concentrations varies from test to test, e.g. depends on the dose of the drug administered.

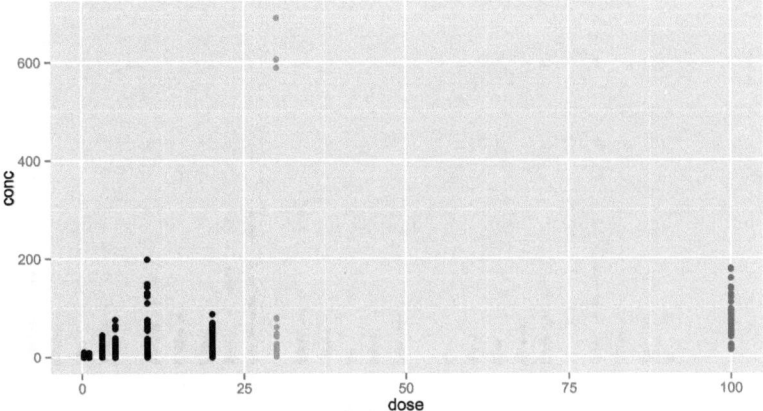

Fig. 2.3 Dependence of concentrations (conc, in total radioactivity) on dose (dose, in mg); test S19 marked in blue, test S2 in orange

We check this hypothesis by considering the dependence of concentrations on the dose administered (see Fig. 2.3). This reveals that in test S2 (marked in orange) the dose of the drug administered is clearly highest (100 mg), followed by test S19 showing the second largest dose (30 mg). Therefore, if we want to compare the results of the different tests, we have to normalize our data. This is the subject of the following analytical step.

2.3.3   Data Preparation

Only rarely can the raw data be used the way they were collected. Therefore, data preparation includes steps such as data cleansing, a first variable selection, and the transformation of variables.

From the descriptive analysis we have concluded that normalization of the data is indispensable. One way of data normalization would be to measure the concentration of the radioactive material in %, i.e. in relation to the quantity of the drug administered. However, we choose a different approach. We divide the concentrations in all organs by the concentration in blood (O1). After normalization of the data, the distribution of relative concentrations looks quite different (compare Figs. 2.2 and 2.4). The highest concentrations of the drugs are now found in the lung (O5) in tests S4 and S10.

We are interested in the relationship between therapy class T (see Table 2.1), which results from domain knowledge, and the distribution data, which was collected from a test animal. As characteristic features we use the 19 normalized concentration values in the organs O2 (plasma) to O20 (pancreas) (see Table 2.2). The values of O1 (blood) are constant = 1 and the digestive organs are omitted,

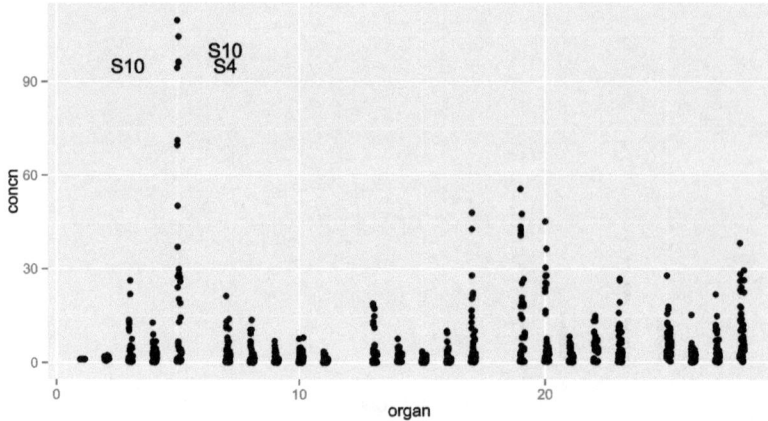

Fig. 2.4 Data after normalization: organ number (organ) versus normalized concentration (concn) (relative total radioactivity); high values in the lung (organ 5) are marked by their test labels (S4, S10)

Table 2.2 Extract from the data set after preprocessing

	Test	Laboratory animal	O2	O3	...	O20	T
1	1	20	1.8713968	0.4354763	...	0.2342368	7
2	1	21	1.8691248	0.5065995	...	0.2586495	7
3	1	22	1.9026714	0.5452909	...	0.2648775	7
⋮	⋮	⋮	⋮	⋮	⋮	⋮	⋮
12	4	29	0.8369208	13.7023740	...	10.3868690	2
13	4	30	0.7531093	13.0646621	...	2.2341045	2
14	4	31	0.7399002	11.1835831	...	3.0313822	2
15	4	32	0.8423113	10.8064324	...	8.5449429	2
⋮	⋮	⋮	⋮	⋮	⋮	⋮	⋮

because the effect of the drugs on these organs is impaired due to early excretion and since the effect on digestive organs is of no interest for our therapeutic classes. We exclude some further observations, which were of no use to the study, e.g. for which less than 20 concentration values are available. This leaves us with 79 observations (laboratory animals) for data analysis and evaluation.

2.3.4 Modeling

The next step of CRISP-DM is modeling, which is the most important step of any statistical analysis. It consists of the application of various modeling techniques, including optimal estimation of its parameters.

In the following, we will learn rules for the relationship between therapy classes 2, 3, and 7 and the features O2, ..., O20. Class 7 includes all other therapy classes. We focus on classes 2 and 3 because of the relatively clear assignment of target organs to classes 2 (brain) and 3 (heart). Class 2 comprises 17 observations (laboratory animals), class 3 has 11 observations, and the residue class 7 is the largest class with 51 observations. The features 'test' and 'laboratory animal', which only identify the animals, are, of course, dropped from the classification.

For the classification, we use a method in which the original features themselves are used to determine the classes, the so-called decision tree, which generates simple to interpret "If - then" rules. In our example, this method identifies only the features heart (O3) and brain (O11) as important for class prediction and the following assignment rules:

Rule 1:   If O3 < 2.1, then select class 7.
Rule 2:   If O3 ≥ 2.1 and O11 < 1.4, then select class 3.
Rule 3:   If O3 ≥ 2.1 and O11 ≥ 1.4, then select class 2.

The only rule that is not 100% accurate is rule 1, because the condition O3 < 2.1 also applies to four observations (laboratory animals) of class 3.

The results can be represented by means of a scatter plot in the variables O3 and O11 (see Fig. 2.5). It can be seen that the three areas segmented by the dotted lines, which correspond to the splits found by the classification method, can be assigned relatively clearly to the classes. The top-right area is assigned to the antidepressants

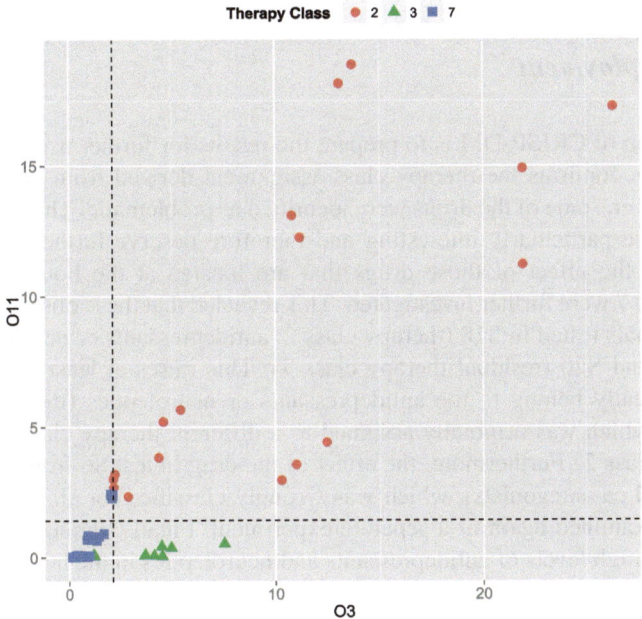

Fig. 2.5 Partition of feature space by the decision tree

and neuroleptics (class 2, red dots), the bottom-right area to the beta blockers and ca-antagonists (class 3, green triangles), and the left area to the residue class 7 (blue squares). The misclassified observations of class 3 (beta blockers and ca-antagonists) are partially obscured by blue squares in the lower left.

2.3.5   Evaluation

The statistical model identifies a partition of the feature space. This partition then needs to be evaluated and interpreted. In particular, it should be assessed whether all objectives of the analysis have been achieved.

The assignments of classes 2 (antidepressants, neuroleptics) and 3 (beta blockers, ca-antagonists) clearly make sense. Class 2 drugs show a high concentration in the brain (O11) and class 3 drugs a high concentration in the heart (O3). Note the low concentration of antihypertensives in the brain and of the remaining drugs both in the heart and the brain. The error rate of 5%, i.e. 4 errors in 79 observations, is acceptable. However, the feature space presented in Fig. 2.5 also shows a somewhat blurred boundary and thus problematic overlap between the antidepressants and neuroleptics (class 2) and the residual class 7, misclassified observations of the beta blockers and ca-antagonists (class 3), and sometimes very high concentrations of the antidepressants and neuroleptics in the heart. This will be discussed in the next step.

2.3.6   Deployment

The final step of CRISP-DM is to prepare the results for further use. All in all, the data analysis confirms the therapy class assignment derived from domain knowledge. However, some of the drugs were identified as problematic. These are the ones that might be particularly interesting and therefore deserve further investigation. First of all, the effect of those drugs that are located at the boundary between classes 2 and 7 were further investigated. This revealed that these observations relate only to animals tested in S18 (therapy class 2: antidepressants or neuroleptics) (see Table 2.1) and S20 (residual therapy class 7). This raises at least two questions: Does S18 really belong to the antidepressants or neuroleptics (therapy class 2)? Does S20, which was originally assigned to a different therapy class, also belong to therapy class 2? Furthermore, the effect of the drug in test S6 from class 3 (beta blockers and ca-antagonists), which was wrongly classified for all 4 observations, should be examined again in a separate experiment. Finally, it should be clarified whether the high levels of antidepressants and neuroleptics in the heart make sense.

2.4   Further Reading

The remarks on the pre-clinical phase are translated from the internet publication https://www.celgene.de/wie-entsteht-ein-medikament-entwicklung-und-studien/.

For CRISP-DM, the book "Data Mining for Dummies" by M.S. Brown (2014) is recommended. The fact that CRISP-DM is still used intensively today is confirmed, e.g., by the internet platform KD-Nuggets: https://www.kdnuggets.com/2014/10/crisp-dm-top-methodology-analytics-data-mining-data-science-projects.html.

The data used here are taken from earlier work conducted by the author at CIBA-Geigy (Switzerland) and were anonymized and re-evaluated for this chapter.

More details on decision trees, e.g., can be found in the book "An Introduction to Statistical Learning" by G. James et al. (2017).

Chapter 3
Drug Studies: Using Statistics to Achieve the Optimal Dose

Holger Dette and Kirsten Schorning

Abstract Only few potential drugs pass clinical trials and become approved. With better statistical methods for determining the optimal dose, the number of successful substances could be increased.

3.1 The Three Clinical Test Phases

Potential new drugs undergo rigorous testing before they receive approval. Most fail and only 0.01–0.02% come on the market. However, some candidates are wrongly rejected because pharmaceutical companies do not conduct the final tests on humans with the optimal dosage of the drug. Together with the biostatistics department of Novartis (Switzerland), we have developed a new statistical method to plan dose-finding studies more efficiently.

In the approval of drugs, a distinction is made between three clinical test phases. In phase 1, the substance is tested on humans for the first time. The aim is to find out how well it is tolerated, how it is distributed in the body (cf. Chap. 2), and how the body processes and excretes it. Phase 2 involves studying the effect of the substance and determining the optimum dosage on the basis of trials involving a few hundred patients. In the third phase, the substance is finally tested with the optimal dose determined in phase 2 on several thousand patients over a longer period of time.

However, what happens if the optimal dose is not used in phase 3, but the drug is over- or underdosed? In the first case, the new substance will probably not pass the tests because it triggers too many side effects. On the other hand, if the dose is too weak, the desired effect may fail to materialize.

H. Dette
RU Bochum, Stochastics, Bochum, Germany
e-mail: holger.dette@ruhr-uni-bochum.de

K. Schorning (✉)
TU Dortmund, Department of Statistics, Dortmund, Germany
e-mail: schorning@statistik.tu-dortmund.de

C. Weihs et al. (eds.), *Statistics Today*, Society, Environment and Statistics, https://doi.org/10.1007/978-3-662-68907-3_3

3.2 The Optimization of Phase 2

One of the goals of phase 2 is, first of all, to find the minimum dose required to achieve the necessary effect such as lowering the blood pressure by a certain degree without causing too many side effects. However, the minimum dose is, of course, not the optimal dose. How can we find it then? Up until now, participants are divided into several groups of equal size in Phase 2. The potential range of the dose, say 0–150 milligrams, is also divided equally (on a linear or logarithmic scale) so that each group is given a specific dose. For example, the first group receives 0 milligrams of the active ingredient, i.e., a placebo, the second group 30 milligrams, the third 60 milligrams, the fourth 90 milligrams, and so on.

In our view, this approach is suboptimal. With a stronger focus on statistics in the study design, one could determine the optimal dose much more accurately. But how? To do so, we must first understand how the dose and its effect are related in appropriate mathematical models. Figure 3.1 shows two examples of different mathematical models that define the relationship between the dose and effect of a drug. If one knew the model for a substance and all its relevant parameters (a, b, c), one could easily determine which dose (MED, minimal effective dose) one must use to achieve the specific desired effect (in this example, an effect of 0.6).

Pharmacokinetic research has shown that, in principle, only few functions can be used to describe the entirety of dose-response relationships of drugs. The different types of functions can be determined from chemical reaction equations using the theory of differential equations, a traditional branch of mathematics.

An example is the EMAX model (MAXimum Effect Model):

$$f(x) = a + \frac{bx}{c + x}.$$

Fig. 3.1 Two mathematical models on the relationship between dose x and effect $f(x)$ of a drug

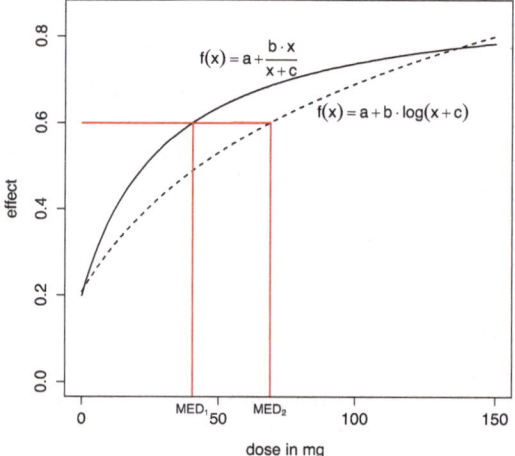

In the formula, the function $f(x)$ assigns a specific effect to each dose value x; a, b, and c are drug-specific parameters. If one knew the dose-response function and the parameters a, b, and c for the new drug, one could easily read off the minimum effective dose from the graph or calculate it using the formula (Fig. 3.1). The problem is you know neither the model nor the parameters when developing a new drug. So, through Phase 2 testing, pharmaceutical companies need to develop good models to describe the dose-response relationship to approach the optimal dose as closely as possible. However, dividing all subjects equally among six dose levels is, to our knowledge, not the best way to do this.

Approaching this in a simplified way, let us assume that the effect of our drug depends linearly on the dose, i.e. we can depict the dependence by a straight line. We gather a set of measured values through which we place a line of best fit (see Fig. 3.2). Figure 3.2a–c show that it is important to choose the measurement points carefully, because:

(a) If one collects data at one x-value only, no single line of best fit can be determined for the measured values. It could be placed anywhere in the range of the shaded area. The minimum effective dose (MED) could therefore lie anywhere between the red lines.
(b) The measurement at two x-values narrows down the range in which the straight lines may occur. The dispersion (σ) for the MED is much smaller.
(c) The dispersion is further minimized if the two x-values are as far apart as possible.

3.3 It is All About the Experimental Design

As shown in Sect. 3.2, our line of best fit can assume any slope and we cannot make any statement about the minimum effective dose if we measure all values at one dose level only. If we measure at two x-values, the situation looks better. However, if we choose dose levels that are too close to each other, we may still end up with lines of best fit with different slopes. The situation improves if we choose doses that are as different as possible, as shown in Fig. 3.2.

However, the relationship between dose and drug effect is generally not linear at all but actually follows more complex models, e.g. our EMAX model. In addition, we would like to determine how many patients should best be treated with which dose without knowing the exact underlying function. To answer these questions, the following should be taken into consideration: Years of pharmaceutical research have shown that only few models exist to describe the dose-response relationship. Additionally, results from phase 1 of the clinical trials already provide first information about the drug-specific parameters.

Fig. 3.2 Line of best fit for different data situations with observations x. The shaded area shows the possible variation of the line of best fit. MED indicates the variation of the minimum effective dose

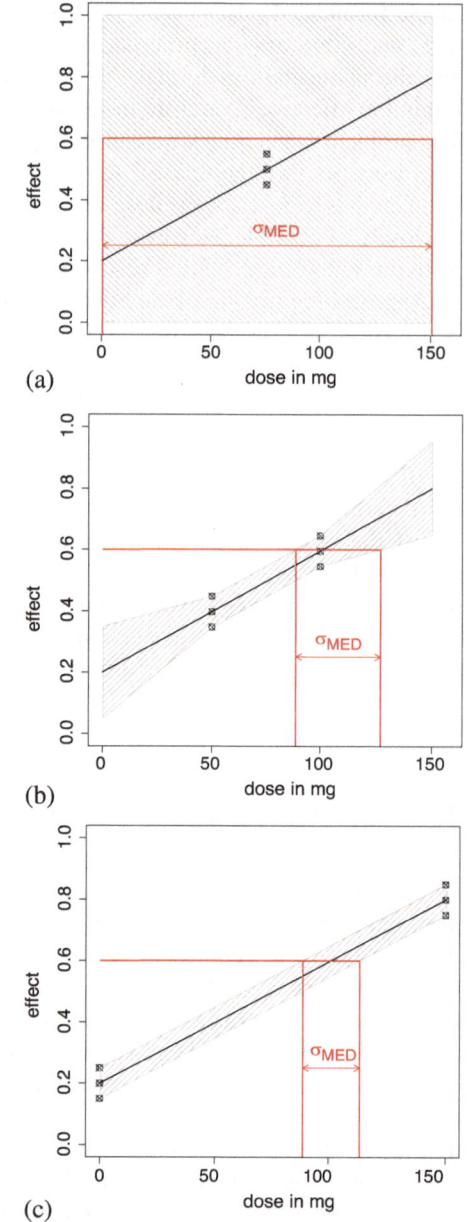

(a)

(b)

(c)

For our method, we essentially have to solve an extreme value problem. In principle, we do the same as in curve sketching as taught in school—only by means of different functions. The functions used in the present study have an important additional property: they are concave (see Fig. 3.3). To determine the maximum

Fig. 3.3 A concave function. The graph also illustrates two secants that lie below the curve (red) and the tangent to the maximum of the concave function at $x = 0.5$ (blue)

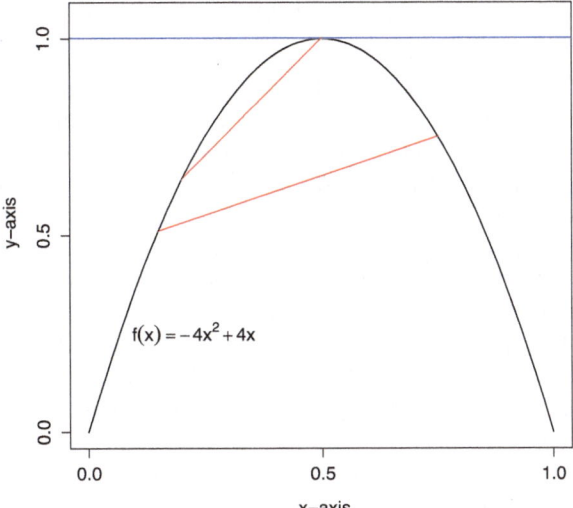

points of a concave function, one sets its first derivative equal to zero. Though we are dealing with more abstract functions than found in school mathematics, the underlying principle is similar. However, in our case we are solving an infinite-dimensional extrema problem.

To that end, we describe a special experimental design with the help of the following matrix:

$$\xi = \begin{pmatrix} d_1 & d_2 & \dots & d_n \\ P_1 & P_2 & \dots & P_n \end{pmatrix}$$

The variable n indicates how many different doses should be tested. $d_1, d_2, \dots, d_n$ denote the doses to be calculated and $P_1, P_2, \dots, P_n$ the relative proportions of patients that should receive the respective dose. In order to determine the accuracy of the estimate of the minimum effective dose, we determine a function that describes the accuracy dependent on the experimental design, i.e. the choice of d and P. Substituting numerical values for the variables—for example, 0 milligrams of drug for one-eighth of the patients, 80 milligrams of drug for one-fourth, and so on—we obtain a value that indicates how accurately we could estimate the optimal dose by means of the experimental design. By maximizing this function, we then determine the best possible experimental design. The statistical challenge is that one cannot specify an upper bound for the number of potential doses n. This makes the extreme value problem infinite-dimensional.

3.4 Towards Practical Application

The new method could be used to obtain a more accurate estimate of the minimum effective dose in phase 2 of the clinical trials than the method used so far, but has not been put into practice yet. Two reasons can be identified for this: The first is of a practical nature. As part of the method used so far, the test subjects might take, say, two, three, or seven pills. Our method might suggest that they should take 3.78 pills. For obvious reasons, this is not feasible. Therefore, we had to extend our method so that the outcome is restricted to integers. Secondly, it takes time for an innovation to gain acceptance since it is never easy to convince clinicians to abandon the approved standards. If, for example, we said that to find the minimum effective dose they should only test three different dose levels when applying our EMAX model, they would reply that our results might be incorrect since we are not using the right model. This is why we have extended our procedure in a way that it works for a large number of models and will not fail simply because one selects the wrong dose-response curve.

Nevertheless skepticism must first be overcome. After all, if something went wrong, this might cause exorbitant costs. What is more, it is not the clinician who ultimately decides whether or not the drug will be introduced onto the market, but an authority. In the U.S., this would be the 'Federal Drug Administration' (FDA). If we are lucky, the FDA will one day adopt our method. Then, doctors could implement it without fearing that the agency might not accept their clinical tests.

3.5 Further Reading

The field of optimal experimental design has been a widely researched subfield of mathematical statistics for decades; see, for example, the monograph "Optimal Design of Experiments," 2006, SIAM, authored by Friedrich Pukelsheim, for an overview. Our own results can be found, among others, in "Optimal designs for comparing curves," Annals of Statistics 44(3), 2016, 1103–1130, and in the article "Optimal designs for active controlled dose-finding trials with efficacy-toxicity outcomes," co-authored by the present authors and Katrin Kettelhake, Weng Kee Wong, and Frank Bretz, published in Biometrika 104(4), 2017, 1003–1010.

Chapter 4
Statistical Alarm Systems in Intensive Care Medicine

Roland Fried, Ursula Gather, and Michael Imhoff

Abstract Monitoring critically ill patients by means of modern computer and measurement technologies generates great amounts of data. The real time statistical analysis of these data provides important insights into patients' state of health and the necessary medical measures needed to save lives and avoid false alarms.

4.1 Alarms in Acute Medical Care

The rapid identification of a patient's critical condition can save lives in intensive care units, operating rooms, emergency rooms, and other acute care settings. However, hospital staff can hardly keep an eye on all acutely endangered patients at the same time. Therefore, their vital functions are measured electronically and monitored automatically (Fig. 4.1). If these values change critically, the system sets off an alarm. Among the multitude of characteristics traced by the system, cardiovascular variables such as heart rate, blood oxygen saturation, temperature, central venous as well as upper (systolic), lower (diastolic), and mean arterial and pulmonary arterial blood pressure are of particular importance for diagnosing circulatory failure (Fig. 4.2). Multiple relationships and interactions exist between these characteristics. For example, high values of one blood pressure type in combination with low values of another type may indicate different forms of circulatory failure. This is why such characteristics have to be jointly assessed.

R. Fried (✉) · U. Gather
TU Dortmund, Department of Statistics, Dortmund, Germany
e-mail: msnat@statistik.tu-dortmund.de; ursula.gather@tu-dortmund.de

M. Imhoff
RU Bochum, Faculty of Medicine, Bochum, Germany
e-mail: mike@imhoff.de

C. Weihs et al. (eds.), *Statistics Today*, Society, Environment and Statistics,
https://doi.org/10.1007/978-3-662-68907-3_4

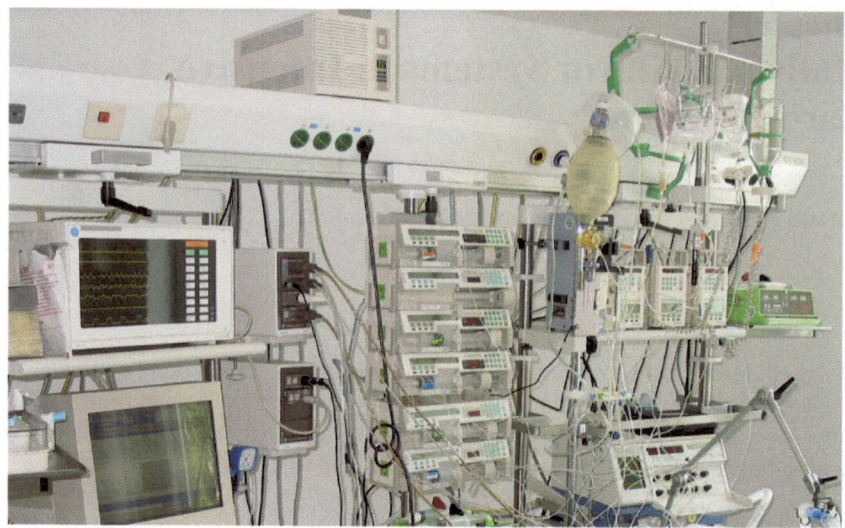

Fig. 4.1 Standard measuring equipment for monitoring patients in intensive care units in larger German hospitals (© own photo. All Rights Reserved)

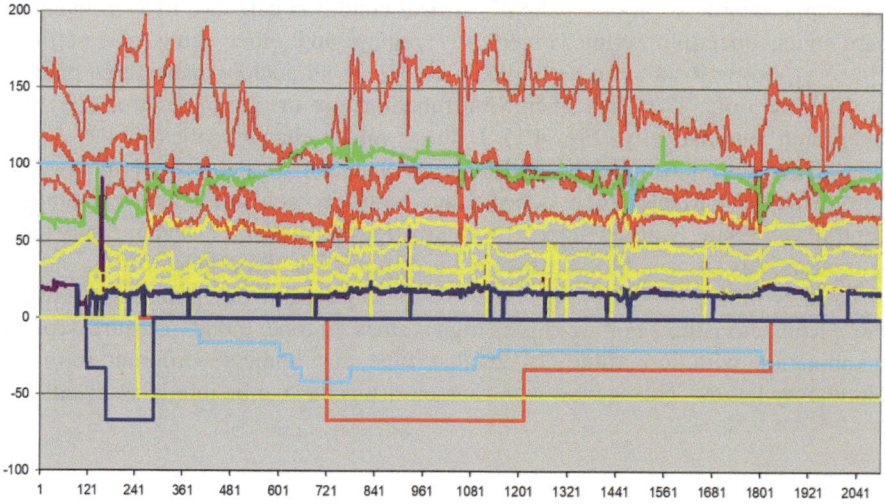

Fig. 4.2 Pressures and oxygen saturations in different vascular regions as well as drug dosages depicted as continuous infusion rates of strong cardiovascular drugs (mirrored at the zero line) for an intensive care patient; observation period 35 hours (2 100 minutes)

Conventional alarm systems are based on the individual upper and lower thresholds for each of these characteristics. As soon as a value exceeds or falls below such a threshold, an alarm is triggered. Thresholds are manually adjusted for each patient, taking into account age, gender, and his/her general and acute physical condition. For example, a heart rate below 50 bpm is alarming for many people, but not so for people who are very active in sports.

Some characteristics are measured every second, with a variety of confounders blurring the readings. Patients' movements, for example, often cause clinically irrelevant measurements. Some of these affect several, consecutive observations and trigger numerous false alarms in combination with other factors. Existing alarm systems show a high sensitivity for detecting critical conditions but lead to high numbers of clinically irrelevant alarms. A number of studies have assessed the rate of false alarms produced by conventional systems to be as high as 90%. This leads to high stress levels and the desensitization of hospital staff. Therefore, conventional systems allow for alarms being set in such a way that they are only triggered after threshold values have been exceeded several times or for a longer period of time. However, this only partially eliminates false alarms. If alarms are triggered for multiple patients at the same time, simple systems provide little information about which alarm is more urgent. Therefore, the high number of false alarms is not only a stress factor for the hospital staff, but also a risk factor for the patients. If, as a result of the high number of alarms, a critical condition remains unnoticed and is treated too late, this can have dramatic consequences for the patients.

In addition to false alarms, making the right diagnosis under time pressure can lead to further problems. The human capacity to absorb and process information is limited. Psychological studies have shown that people can normally take in no more than seven pieces of information at a time and have problems establishing connections between more than two of them. In intensive-care medicine, however, medical practitioners are confronted with high-frequency measurements of about a dozen cardiovascular variables and other medical characteristics for several critically ill patients at the same time. This shows the urgent need for intelligent alarm and analysis systems that allow for reliable decision-making under stress.

In the following sections, we outline some statistical methods that we have explored for the development of intelligent alarm and analysis systems in an interdisciplinary research project. In addition to real-time capability and reliability, comprehensibility of such monitoring systems is important for the medical professionals who ultimately make the decisions and thus need to understand the alarms and recommendations of the system.

4.2 Smoothing as Part of Data Preprocessing

One way to refine patient monitoring in order to reduce the number of false alarms and to obtain more reliable information about the patient's condition is to improve the traditional system of thresholds. One way to do this is to preprocess the data to mask out irrelevant events from patients' movements and other short-term fluctuations of characteristics. To this end, we have developed robust smoothing techniques that allow a 'denoising' of the data and reduce them to the clinically relevant information. This included a thorough investigation of the strengths and weaknesses of the many smoothing methods proposed in the literature with regard to our intensive care context.

Ultimately, the decision was made to use filtering methods based on sliding time windows. Filtering is broadly applicable and requires weak assumptions on the underlying signal and noise only. The evaluation of the patient's current condition is, for example, based on measurements of the time window of the last half hour. The best-known method to evaluate such time windows is the 'moving average', i.e. the arithmetic mean of the measurements in the time window. However, the result would be strongly influenced by measurement outliers. Instead, we employ a 'moving median' by sorting the data of the time window by size and using the value in the middle.

However, like a moving average, a moving median also lags behind the patient's actual condition by half a window width, e.g. a quarter of an hour, if the data, e.g. measurements of blood pressure, show an upward or downward trend, for example in the wake-up and stabilization phase immediately after surgery. These two approaches assume that the level of the measured values is almost constant in the time window and are therefore inadequate to predict such trends. To mitigate the time lag, one can linearly model the trend in the time window, estimate it by means of regression, and use the estimated value of this trend to determine the value of the characteristic at the current point in time. To make sure that noise has little to no effect on the fit, we use robust regression methods with similar properties as the median, but not the classical least squares method. To select and refine a suitable method, we have conducted extensive comparative studies with real and simulated data to find a particularly well-suited method.

In addition, rules for the choice of the window width are needed. Is the patient's current condition best characterized by the data of the last half hour, or is it better to use the data of the last 15 minutes only, or rather the data of the last full hour? The assumption of a linear progression, which justifies fitting a straight line, is at best locally valid. If one chooses too long a time window, one commits a systematic error, distorting the estimates. On the other hand, if the time window is too short, one might waste potentially useful information. In a longer hospital stay, both quiet periods and periods with greater variation in patients' conditions exist. Therefore, adaptive methods that continually adjust the window width to the progression of medical characteristics are particularly promising. Of course, this needs to be automated, without constant manual readjustment needed. Therefore, we have developed algorithms for estimating the signal values, consisting of precisely described sequences of calculation and verification steps.

The alarm can then rely on data that have been denoised by robust smoothing techniques. Furthermore, the thresholds at which the alarm is triggered can be closer than those that rely on the much noisier original data. In Fig. 4.3, for example, the dashed thresholds result in alarms at minutes 955, 1271, and 1555 when the smoothed signal is used. A conventional alarm system, on the other hand, monitors the unsmoothed measurements and would produce 33 alarms for the same thresholds, even if alarms were not triggered until thresholds are exceeded at least twice. Thus, the robust system provides more accurate information and the number

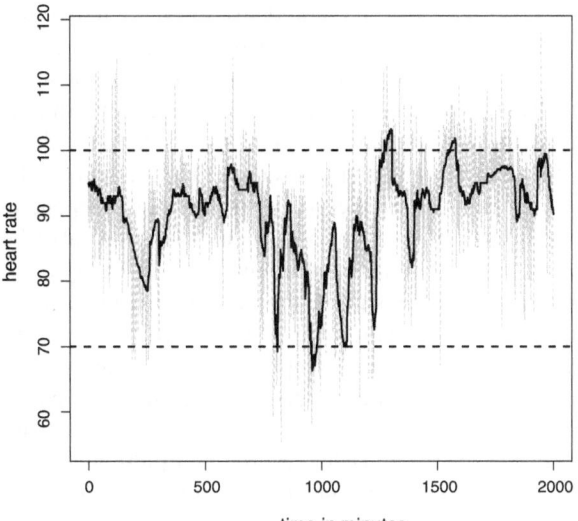

Fig. 4.3 Heart rate measurements of an intensive care patient (gray) during 2000 minutes and a signal extracted by means of an adaptive robust regression filter in real time (black); thresholds dashed

of false alarms decreases. Threshold alarms can also be improved through, e.g., predictions about the future development of the characteristics by means of the estimated slope of the fitted line. This visualizes the recent development of the patient's condition and helps predicting the near future and thus detecting potentially dangerous developments at an early stage.

4.3   Joint Analysis of Characteristics

The smoothing methods described above analyze the development of a single characteristic over time. However, a number of interactions exist between the collected characteristics, so that one characteristic may also contain information about others. Therefore, a joint analysis with multivariate methods is worthwhile. For such an analysis, the previously mentioned limitations of human information processing should also be considered. Thus, it makes sense to compress the many observed characteristics into a manageable number that is as meaningful as possible.

Such a compression can be achieved through a well-conceived selection of complementary characteristics with as little redundancy as possible. For this purpose, an overview of the relationships between the characteristics is useful. A correlation analysis provides first insights. The ordinary correlation coefficient between two characteristics measures the strength of a linear relationship on a scale from -1 to $+1$. If the value of this coefficient is (close to) 0, the characteristics are uncorrelated, i.e. there is at least no linear relationship between them. If the value is $+1$ or -1, a strong correlation exists, i.e. there is an exact linear relationship, which is either increasing or decreasing, respectively. However, even if the absolute value of the correlation is +1, one should be careful not to overinterpret this as a direct influence of the one specific characteristic on the other, in particular not as a causal effect. Often, correlations also arise from relationships with other characteristics.

Therefore, it is worth looking not only into ordinary correlations between characteristics but also into partial correlations. The latter are those correlations that remain after one has removed the linear effects of further characteristics by means of regression. Partial correlations obviously depend on which further characteristics are taken into account and should therefore also be interpreted with caution. To analyze the correlation structure between many characteristics, so-called graphical models, which visualize the correlations by means of a dependency graph, have been developed since the 1970s based on a profound mathematical theory. Our work shows that such graphical models can be successfully used for information retrieval even for dependencies in an intensive care context that have a potentially delayed discovery. In Fig. 4.4, for example, strong correlations exist primarily between heart rate (HR), arterial and pulmonary arterial blood pressure (ABP and PABP, respectively), and PABP and central venous blood pressure (CVBP). The observed relationship between oxygen saturation (SpO2) and temperature (Temp) is a measurement artifact because the instrument measuring SpO2 responds in a temperature-dependent manner.

Selecting particularly informative characteristics is not the only way to compress information. Another approach is principal component analysis. Joint observations of p characteristics are, from an abstract perspective, points in a p-dimensional space. Principal component analysis looks for directions of maximal variation of

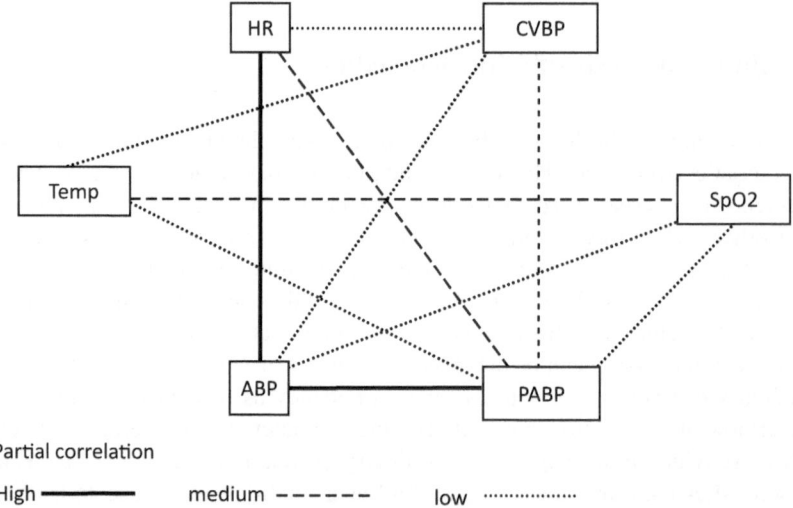

Fig. 4.4 Graphical model based on partial correlations for the cardiovascular variables heart rate (HR), arterial and pulmonary arterial blood pressure (ABP), and (PABP) as well as central venous blood pressure (CVBP), oxygen saturation (SpO2), and temperature (Temp)

Fig. 4.5 Central venous (green) and pulmonary arterial blood pressures (systolic red, diastolic blue, and mean pressure magenta) of a patient and the principal component extracted from these four characteristics (black)

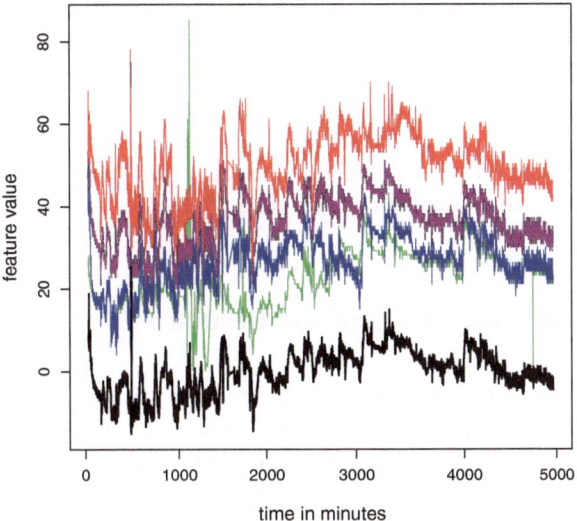

the p-dimensional observations of the characteristics. The idea is to use as few directions of this kind as possible and lose as little variation as possible.[1] Standard rules for choosing the number of principal components are usually based on the variability explained by the principal components and the remaining residual variability. The interpretation of the directions can be improved by rotation to obtain another set of vectors spanning the same space. Our analyses have shown that rotations generating interpretable (artificial) new features from the original cardiovascular characteristics can retain more relevant information about the original features than feature selection. In addition, the new features determined by a principal component analysis are uncorrelated with each other, so it is more adequate to analyze these features separately than the original characteristics. Figure 4.5, for example, shows that a principal component provides the essential information in a compressed and less noisy way and fluctuates around 0 as the result of centering. In addition, positive values of the principal component can be interpreted as high, while negative values can be interpreted as low, regardless of the patient.

In addition to the methods described in this chapter, other dimension-reducing methods exist. While the principal component analysis yields uncorrelated features without linear dependencies, the independent component analysis aims to create independent features without nonlinear dependencies. This requires stronger assumptions, in particular that not more than one of the artificial features to be

---

[1] Mathematically, this is solved by a 'spectral decomposition' of the matrix of variances and covariances of the features; one determines the so-called eigenvalues and eigenvectors of this matrix. The desired directions are determined by the eigenvectors that have the largest eigenvalues. One can then project the p-dimensional observation points onto these direction vectors and thereby generate new observation values.

extracted is normally (Gaussian) distributed. Another approach to dimension reduction, often applied completely under the Gaussian assumption, is factor analysis. It is used especially by psychologists and increasingly by economists to explain many measurable characteristics by means of few interpretable features (latent factors). In our intensive care context, when using such methods, it is important to ensure that robust estimation methods that disregard the numerous measurement artifacts and are still computable in real time are used.

4.4 Validation of the Results

The statistical solutions identified in the present chapter have to be tested and evaluated prior to clinical application. For validation, a medical professional needs to simultaneously monitor the patient and the system for a while. The results provided by the automatic system are then to be compared with the expert's observations.

However, different experts may come to different assessments, and experiments show that even the same expert may assess the same situation differently at different times. People interpret facts not only against the background of general but also of recent experience. Therefore, even a validated system can only offer a compromise of different expert assessments. In addition, this type of validation is extremely time-consuming and cannot always be performed. Methods of statistical learning can help to ultimately create a computer-based set of rules for expert evaluations of alarm situations.

4.5 Further Reading

Statistical alarm systems have their origins in industrial process control. The pioneer in this field was Walter A. Shewhart with his 1931 book "Economic Control of Quality of Manufactured Product." Since then, alarm systems have been developed for increasingly complex situations considering many new and increasingly complex application problems.

A comprehensive assessment of alarm procedures in critical care medicine is provided in the 2010 VDE position paper "Alarming of Medical Devices." The filtering methods for robust signal extraction from Sect. 4.2 is further explained in Schettlinger, Fried, and Gather (2010): "Real Time Signal Processing by Adaptive Repeated Median Filters," International Journal of Adaptive Control and Signal Processing 24, 346–362. The multivariate analysis approaches described in Sect. 4.3 are discussed in detail in Gather, Imhoff, and Fried (2002): "Graphical Models for Multivariate Time Series from Intensive Care Monitoring," Statistics in Medicine 21, 2685–2701, and Gather, Fried, Lanius, and Imhoff (2001): "Online Monitoring of High Dimensional Physiological Time Series – A Case-Study," Estadística 53,

259–298. Further information on the validation of the results presented in Sect. 4.4 is available in Siebig, Kuhls, Imhoff, Langgartner, Reng, Schölmerich, Gather, and Wrede (2010): "The Collection of Annotated Data in a Clinical Validation Study for Alarm Algorithms in Intensive Care - a Methodologic Framework," Journal of Critical Care 25, 128–135.

Chapter 5
Personalized Medicine: How Statistics Helps Not to Drown in the Flood of Data

Jörg Rahnenführer

Abstract The number of new cancer diagnoses and the number of people dying from cancer are rising continuously in most countries around the world. It is very difficult to find the optimal treatment for a particular cancer case, because cancer is not a homogeneous disease, but many different manifestations exist, all of which are grouped together under the general label 'cancer'. However, statistical methods can help identify groups of patients that are similar in terms of genetic characteristics. For some such groups, a targeted therapy can take into account the genetics of the patients for a more promising treatment.

5.1 Genetic Decision Support in Medicine

Cancer has often been referred to as the scourge of humankind. It is one of the primary causes of death in many countries around the world, with prostate cancer in men, breast cancer in women, and lung cancer and colorectal cancer in both sexes being particularly frequently diagnosed. For many of these cancer types, numerous subtypes exist. This is important for identifying the right treatment. For example, lung cancer is usually caused by smoking, but it can also occur in nonsmokers. Furthermore, it is important for the treatment which tissue is affected, e.g. glandular tissue in adenocarcinomas or the bronchi in bronchial cancer.

Even for patient groups that are homogeneous in terms of their clinical picture, the genetic subtype of the disease plays a major role. Patients with so-called non-small cell lung cancer (NSCLC) are usually tested to determine whether a specific gene mutation can be found in the tumor tissue which produces the protein EGFR (epidermal growth factor receptor). Such mutations are the result of genetic changes, when compared to the prototypical case. Another reason for the NSCLC is an overproduction of the protein EGFR. In both cases, the tumor cells grow

J. Rahnenführer (✉)
TU Dortmund, Department of Statistics, Dortmund, Germany
e-mail: rahnenfuehrer@statistik.tu-dortmund.de

© The Author(s), under exclusive license to Springer-Verlag GmbH, DE, part of Springer Nature 2024
C. Weihs et al. (eds.), *Statistics Today*, Society, Environment and Statistics, https://doi.org/10.1007/978-3-662-68907-3_5

37

uncontrollably and proliferate. For this situation, specific cancer therapies are now available, which can block the signal from EGFR and thus stop tumor growth. For patients without EGFR alterations, conventional chemotherapy has far better chances of success. The genetic status thus helps determine the most appropriate therapy.

5.2   Efficacy and Side Effects of Therapies

The above example shows what is important in personalized therapies. In this respect, patients need to be divided into subgroups, taking into consideration the efficacy and side effects of therapies. Figure 5.1 illustrates the differences between classical and personalized medicine in the reaction to therapy for a group of patients with the same diagnosis. While the green group benefits from the treatment without side effects, the yellow group shows no efficacy but also no (bad) side effects. For the red group, however, side effects are so severe that efficacy of therapy is irrelevant.

Therefore, for successful therapies methods are needed that identify subgroups of patients that belong to the green group shown in Fig. 5.1. Including the individual genetic characteristics of the patient or a tumor and thus the true molecular causes for a disease, as in the EGFR example, promises great potential for better medical decision making.

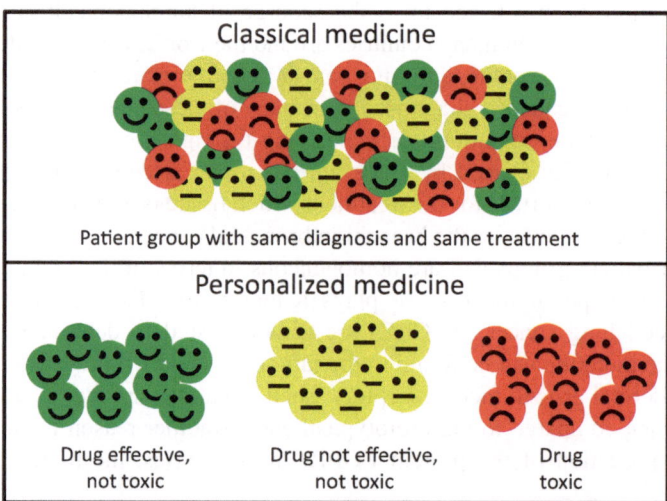

Fig. 5.1 Differentiation of patients according to efficacy and side effects of a medical therapy

5.3   In Search for Genetic Patterns

In recent decades, tremendous progress has been made in three different areas: (1) in molecular biology, especially in understanding the role of genes and proteins in many important processes that occur in the human body; (2) in the development of technologies for simultaneous measurement of tens of thousands or even millions of genetic values; (3) in the development of high performance computers that enable sophisticated computer-based analysis and modeling of such large amounts of data in a viable period of time.

The technologies can be used to measure the gene activity (gene expression) of tens of thousands of genes, the protein activity of tens of thousands of proteins, or the mutation status of millions of sites in the human genome. For each individual patient, this provides a flood of measurements that can be used for diagnosis and therapy selection.

At the same time, however, this gives rise to a major problem known as the 'curse of dimensionality'. Let us consider the following case: A simple yes/no decision is to be made from genetic measurements by predicting, for example, whether a patient will respond to a certain therapy or not. Even if each genetic measurement was binary only (i.e. 0 or 1, yes or no, gene is mutated or not mutated), with 10,000 measurements, $2^{10,000}$ possible genetic combinations would exist for a single patient. This is an incredibly large number with more than 3000 digits, which is many times more than the number of grains of sand on earth.

To predict whether the therapy will work for a particular patient with a specific individual gene combination, we would have to study all possible variants. This means that one has to make $2^{10,000}$ yes/no decisions concerning the patient's response to a therapy. To this end, one should set up at least one experiment for each genetic variant including, say, 100 patients carrying the specific genetic variant. 50 of them would receive the therapy, the other 50 would not (placebo). A subsequent comparison of the responses of these two groups to the therapy or placebo would reveal the overall response to the therapy. All decisions on the response to the therapy for the different genetic variants would be stored in a database. This would then empower the researcher to decide on the potential success of the therapy for a new patient on the basis of the stored data in relation to their specific genetic set up.

To manage this seemingly intractable challenge, a few restrictions on combinations may help. First, not all possible variants exist in reality; many are medically unreasonable. Second, not all measurements are equally important; only the most important ones need to be identified. Third, correlations exist between important measurements. For example, it is often the case that in a combination of two genes, activity measurements are high or low for one gene if they are high or low for the other, respectively, so that one needs only one of them for prediction.

5.4   Statistical Tricks

The reason why statistics is needed to deal with medical decision making from genetic data is also due to the inevitable noise in the measured values. This means that the measurements for most gene variants are subject to variation for essentially two reasons. First of all, most of the time a gene variant is not identical between different patients. For example, each person has unique combinations in their genetic makeup that are not important for selecting medical treatment but might have at least some influence on the patient's reaction to the treatment. Second, experimental variability may cause noise in measurements of gene activity.

The first statistical trick to help decide whether a therapy is successful or not is to look at potentially influential variables, i.e. properties of individual genes, in isolation. This usually eliminates a large number of the variables right away. This approach may also identify a number of variables that are only coincidentally related to the target variable. Predictions based on such variables would be misleading for predicting the therapeutic success for future patients. Therefore, only those variables that are particularly strongly related to the target variable are considered. Statistical methods are used to calculate how strong the correlation needs to be so that one can be relatively sure that it is not just a chance finding due to the large number of variables.

The second statistical trick is to analyze correlations between variables. If two variables are very similar, the one that is superior for prediction will be retained while the other will be neglected. If a group of variables is known to interact in a biological way, for example because they play a common role in immune defense, which is becoming increasingly important for cancer research, then the values of the individual variables can be combined to form a new variable. The advantage of this procedure is that the noise of this new variable is smaller than the noise of the original variables. This property statisticians often refer to as the 'law of large numbers'.

Finally, the third statistical trick is to evaluate not only the predictive accuracy of a model used for analyzing patient data but also the size of the model, i.e. the number of variables used when constructing a predictive model. Of two models that provide equally good predictions, the simpler one, i.e. the one with fewer variables, should be considered. This is often done mathematically by means of introducing a so-called penalty, as described in the following: A model is evaluated by comparing the predictions and the true values of the target variable. This is summarized by a single measure M, where smaller values of M indicate a better prediction. The overall criterion S for model selection is the sum of value M and the number of variables included in the model. Models with the lowest values of this criterion S are considered the most suitable. In case of models with similar predictive value M, this means that the model with the lower number of variables is selected and larger models are penalized. Identifying the right balance between quality and size of the model is a popular strategy in statistics. It has proven extremely useful for handling the large number of variables in genetic medical research.

The successful implementation of these statistical tricks requires statistical knowledge on the one hand and experience in handling such data and sufficient biological-medical knowledge, often based on intensive collaboration between statisticians, biologists and physicians, on the other.

5.5 Medical Application

The use of genetic measurements already plays an important role in medicine, as the EGFR example in bronchial carcinoma has shown (cf. Sects. 5.1 and 5.2). However, for many diseases, treatment decisions are still purely based on immediately assessable numbers of demographic and clinical variables, e.g. the patient's age, tumor size, tumor type, and hormone receptor status in cancer research. For personalized medicine, however, a large number of genetic measurements need to be included, which makes procedures much more complex.

Because of the extremely large number of ways to combine genetic measurements into a prediction, the use of genetic variables leads to a greater risk of selecting irrelevant variables for prediction than the use of clinical variables. Therefore, it is important to compare models based on genetic variables with purely clinical models to determine whether predictive quality could be improved. Finally, one can construct models that use both genetic and clinical variables. Again, statistical methods should be used to show that genetic variables generate additional benefit, i.e. that for new patients the predictions are more precise compared to the predictions of purely clinical models.

In addition, an important advantage of genetic models is that they can contribute to gaining medical knowledge by interpreting those variables most important for prediction from a biological perspective. For example, it is often analyzed which biological processes and molecular functions are influenced by the most important genes or proteins in a study. This, again, may facilitate biologically motivated therapies. Similar approaches are used in developmental biology and toxicology.

5.6 Models for Estimating Disease Progression

Statistical models can support the development of personalized therapies in many ways. So far, we have mainly considered predictive models. However, genetic patterns can also be very useful for accurate and early diagnosis. One example is models that use statistical methods to estimate the sequence of genetically relevant steps for the progression of a particular disease. The assumption here is that genetic changes always occur in a certain order and that one can determine the progression of the disease on the basis of the genetic changes that have already occurred.

For example, human tumors are often linked to typical genetic events, such as changes in the corresponding tumor cells. The identification of characteristic disease

Fig. 5.2 Genetic cancer
progression in meningioma, a
type of brain tumor: Disease
onset = start event 0, genetic
progression (GP) = 0;
probability 32% that loss of
chromosome 22 occurs in
tumor cells (GP = 2);
probability 18% for
additional loss of parts of
chromosome 1 (GP = 6); etc.

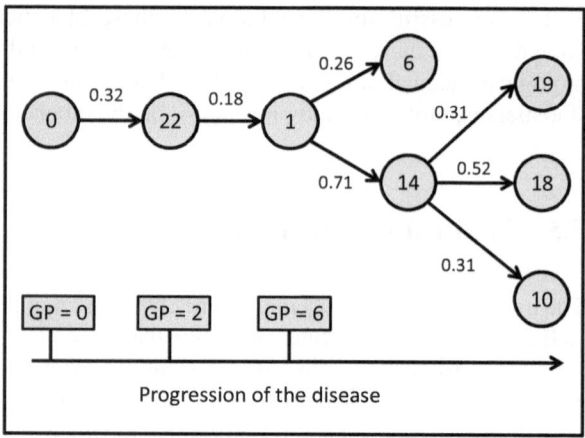

processes in such tumors can then facilitate the prediction of patient survival and thus the selection of the optimal therapy.

In the resulting models, the progress of the disease is typically described by stepwise accumulation of genetic changes in cancer cells. On this basis, the stage of disease progression is determined for individual patients. For example, for various types of brain tumors, a higher genetic progression (GP) score is associated with a shorter time to relapse or death after treatment.

The model in Fig. 5.2 shows genetic steps in meningiomas, a subtype of brain tumor associated with a comparatively good prognosis for patients. For all patients, the progression starts with the start event 0, which represents the onset of the disease. Genetic progression is set to GP = 0. The GP of a tumor is defined as the average 'waiting time' for the genetic aberration to occur. Tumors that are characterized by a low number of aberrations in an early stage receive low GP values, tumors with many late aberrations are associated with high GP values. The waiting time is approximated by the value of $(1-p)/p$ with p = aberration probability at the current stage of tumor development. For a randomly selected patient suffering from the cancer type meningioma, the probability is $p = 32\%$ that a loss of chromosome 22 is observed in the cells of the tumor tissue (cf. Fig. 5.2). The GP for this aberration equals 2, since $(1 - p)/p = 0.68/0.32 \approx 2$. For such patients, the probability of an additional loss of parts of chromosome 1 is about $p = 18\%$, so this event is rarer and can be converted to genetic progress GP = 6 in a statistical model, since $(1-p)/p = 0.82/0.18 \approx 4$ has to be added to GP = 2. After that, further chromosomal changes with the given probabilities can also occur partly independently of each other (6 and 14 as well as 19, 18 and 10), which leads to even higher values of GP.

It is remarkable that this progression model can be generated from so-called cross-sectional data, which, for each patient, is restricted to the data at the time of diagnosis before surgery and do not include follow-up measurements. The progression of the disease can be estimated by a statistical model that assumes the same course for all patients and associates the less frequently observed events with more advanced disease stages.

The medical relevance of this model becomes obvious when patients are grouped by progression status and when their relapse rates are examined. For patients with GP = 6, the number of patients with relapse of meningioma within a certain time after surgery is significantly increased. As a medical consequence, these patients need to be subjected to more regular follow-up examinations than patients with a lower GP.

5.7   Summary

Human life expectancy has been increasing worldwide for more than 100 years. As a result, the proportion of people who get ill and die from genetically based diseases has also gone up; amongst these, cancer can be found particularly frequently. Today, millions of genetic variables can be measured in tumor tissue or in the blood of patients. The measurements usually come with different levels of statistical uncertainty. Statistical methods are therefore a useful and indispensable tool for identifying the most important variables among the countless ones available for medical examination. In personalized medicine, this can help to improve diagnoses and therapy decisions with regards to the patient's genetic status.

5.8   Further Reading

There are countless publications on personalized medicine. The model illustrated in Fig. 5.2, Sect. 5.6, in a simplified way has been published and interpreted in the article by Ketter et al. (2007): "Application of oncogenetic trees mixtures as a biostatistical model of the clonal cytogenetic evolution of meningiomas" in the International Journal of Cancer 121(7), 1473–1480.

Chapter 6
Modulating Genetic Effects on Bladder Cancer Risk in an Area of Coal, Iron, and Steel Industries

Silvia Selinski, Katja Ickstadt, and Klaus Golka

Abstract For many years, genetic data have been collected in addition to personal characteristics such as smoking behavior and occupation when analyzing the causes of disease. Using these different types of data, we investigate whether the structural change in the Ruhr area is reflected in studies on urinary bladder cancer. The Ruhr area is an industrial region in the western part of Germany, which has developed from coal, iron, and steel towards low-emission industries.

6.1 Environment, Genes, and Urinary Bladder Cancer

The Ruhr area is located in the federal state of North Rhine-Westphalia and is the fifth largest conurbation in Europe with 5.1 million inhabitants. The cities of Duisburg, Essen, Bochum, and Dortmund, each with a population between 400,000 and 600,000 inhabitants, constitute its largest cities. For many years, the Ruhr area was the center of the German coal, iron, and steel industries, but their importance has declined steadily since the mid-1960s.

Structural change in the Ruhr area is evident. For example, for the Phoenix steel industry site in Dortmund (Fig. 6.1), we see how air pollution and workplace conditions have changed drastically within just a few decades. Do these environmental changes also affect genetic risk factors? Such interactions between genetics—as individual 'internal' and unchangeable risk factors—and environment—as 'exter-

S. Selinski
IfADo, Toxicology/Systemtoxicology, Dortmund, Germany
e-mail: selinski@ifado.de

K. Ickstadt (✉)
TU Dortmund, Department of Statistics, Dortmund, Germany
e-mail: ickstadt@statistik.tu-dortmund.de

K. Golka
IfADo, Clinical Occupational Medicine, Dortmund, Germany
e-mail: golka@ifado.de

© The Author(s), under exclusive license to Springer-Verlag GmbH, DE,
part of Springer Nature 2024
C. Weihs et al. (eds.), *Statistics Today*, Society, Environment and Statistics,
https://doi.org/10.1007/978-3-662-68907-3_6

45

Fig. 6.1 Site of the Phoenix-East steel industry in Dortmund-Hörde in July 2000 with the Hörder Fackel (flare; left) and in August 2014 with the Phoenix lake (right) (sources: picture left: https://commons.wikimedia.org/wiki/File:2000-07 Hoesch Phoenix-ost.jpg, Bassaar [CC BYSA 3.0 (https://creativecommons.org/licenses/by-sa/3.0)]; picture right: with kind approval by ©Oskar C. Neubauer [2022]. All Rights reserved)

nal' and changeable risk—are of particular interest for the case of urinary bladder cancer. This disease is caused by pollutants that have occurred at many workplaces in the Ruhr area. Metabolism of the pollutants is modulated by genetically modified enzymes. In this context, urinary bladder cancer represents a model for investigating the so-called gene-environment interaction. In patients who are reasonably exposed to such pollutants via their occupation, bladder cancer is considered as an occupational disease (called BK1301 in the German context). However, it is only recognized as an occupational disease if people have been exposed to the pollutants in their daily working lives over years and at a certain exposure level. The recognition as an occupational disease also depends on an expert opinion on the genetic make-up of the patients, e.g., the individually different ability to detoxify the carcinogenic pollutants. Two important genes, i.e. *glutathione S-transferase μ1 (GSTM1)* and *N-acetyltransferase 2 (NAT2)*, increase the risk of urinary bladder cancer in parts of the population or play no role at all, depending on environmental and workplace conditions. Both genes produce enzymes involved in the detoxification of bladder cancer-causing chemicals. If the enzyme is present in lower quantities (for *NAT2*) or is even missing completely (for *GSTM1*), the pollutant can be detoxified in lower quantities. As a result, more pollutants enter the urinary bladder and cause damage to the mucous membrane.

In order to find further risk factors and to examine the known ones more closely, we determine potential influencing factors in different populations, e.g. urinary bladder cancer patients in a clinic, patients without bladder cancer, and suggested occupationally diseased patients (evaluated for BK1301). Such influencing factors include: occupation held, exposure to pollutants, place of residence, lifestyle factors such as smoking behavior, and genetic data. We compare the frequency of these factors in the three groups. The data come from urinary bladder cancer cases in Dortmund in the years 1992–1995, i.e. when pollution levels were still very high, and since 2009, i.e. after the closure of the coal, iron and steel industries. For both periods of investigation, data were also collected from a control group, i.e.

patients without bladder cancer. As illustrated in the following, we could determine astonishing results for the influence of *NAT2* and *GSTM1* on the development of urinal bladder cancer.

6.2   Epidemiology and Genetics

'Epidemiology' studies the distribution and causes of diseases in a population. The beginnings of systematic analyses date back to the mid-nineteenth century. The most prominent examples include John Snow's work on the 1854 cholera epidemic in London, Florence Nightingale's statistical analyses of causes of death in British barracks and hospitals, and Semmelweis' initially controversial work on the causes of childbed fever in 1847/1848. The association of bladder cancer with occupational exposure to certain dyes, initially fuchsine, was first described in 1895 by the German surgeon Ludwig Rehn. Before that (1856–1859), the British chemist and industrialist W.H. Perkin had already reported a frequent occurrence of urinary bladder cancer cases in the paint industry.

Epidemiology generally investigates a possible connection between an influencing variable (risk factor) and a disease in a population on the basis of a study group. To detect a noticeable accumulation of the disease in people with a risk factor compared to people without this risk factor, a whole range of statistical methods exists.

Two main types of studies are cohort studies and case-control studies. 'Cohort studies' examine the ratio of diseased to non-diseased persons in a representative part of the population. They compare the diseased-healthy ratio between the subgroups with (risk group) and without a particular risk factor. Through this, the probability of the disease for the risk group can be calculated in comparison to the population without the risk factor. In a 'case-control study', diseased persons (cases) and matching non-diseased persons (controls) are collected. The comparison group should be as similar as possible to the case group with regard to possible influencing factors that are not of interest to the study but could possibly influence its results (e.g., age and gender). The ratio of the number of persons with and without risk factors is compared between the case group and the control group. In contrast to the cohort study, we can observe the frequencies of the risk factor in cases and controls and can then infer the probabilities of the disease under certain conditions.

In both types of study, it is of interest whether an increased probability of disease really exists in the presence of the risk factor and, if so, the level of risk. The most important measure for this is the 'odds ratio' (OR). It is defined as the ratio between the odds of getting the disease if the risk factor is present and the odds of getting the disease if the risk factor is not present. The risk factor can be a certain genotype or the person's gender, smoking behavior, or occupation. If this odds ratio is greater than 1, the odds of getting the disease increases if the risk factor is present. Whether this increase is random or statistically significant, can be checked with a simple statistical test, the 'chi-square test'. This test compares the squared difference

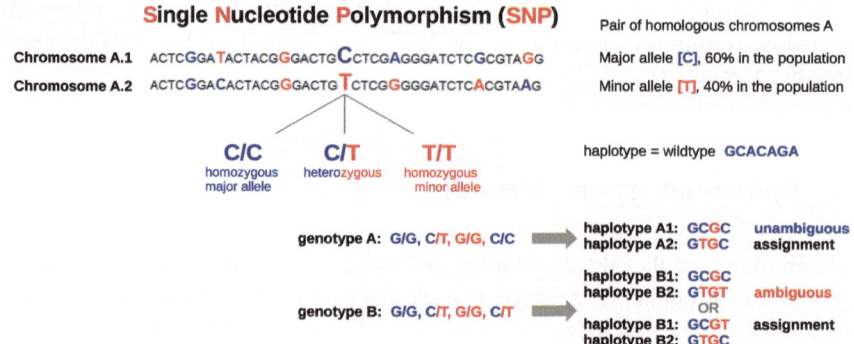

Fig. 6.2 Basic genetic concepts based on an example of a DNA sequence of a homologous pair of chromosomes. C and T denote the bases cytosine and thymine

between the observed frequencies and the frequencies that would be expected if there was no connection between the risk factor and the disease (independence). If the differences are large, this indicates a significant association.

'Genetics' deals with the transmission of hereditary dispositions—or genes—from one generation to the next, as well as the laws governing the expression of these dispositions. Of particular interest is the probability with which the genetic information—the genotype—can be inferred from its appearance—the phenotype.

The genetic information is stored in the DNA (*deoxyribonucleic acid*), a molecule composed of two complementary chains that usually form a double helix. The chains consist of a sequence of four different nucleotides, the organic bases adenine (A), thymine (T), guanine (G), and cytosine (C) (see Fig. 6.2). These bases form the genetic code, so that if their sequence on the DNA strand is known, the genetic information can be inferred. The majority of the DNA is located in the cell nucleus in compact form as chromosomes. These exist as a double set, i.e. half a set of chromosomes (23 in humans) is passed on from each parent to the offspring. These, in turn, have a double set of chromosomes.

If we consider a certain genetic information in a specific position of the DNA (see Fig. 6.2), the bases can be the same on both chromosomes (A.1 and A.2), i.e. are homozygous (e.g., C/C or T/T) or different, i.e. heterozygous (e.g., C/T). One can distinguish between the more frequent homozygous variant (C/C) or the rarer one (T/T). This information at a specific point of the DNA is also called the genotype. If the difference between the bases in the heterozygous case affects a single nucleotide only, it is also called *single nucleotide polymorphism* or *SNP* (pronounced snip). This is the most common type of genetic variant. The different variants at one position, in our case [C] or [T], are also called alleles.

In addition to differences that affect individual nucleotides, variants also encompass entire DNA Sections that differ or are missing. For example, the gene of *glutathione S-transferase μ1 (GSTM1)* is completely missing on both chromosomes (*GSTM1* negative) in about 50% of the European population. This means that half of

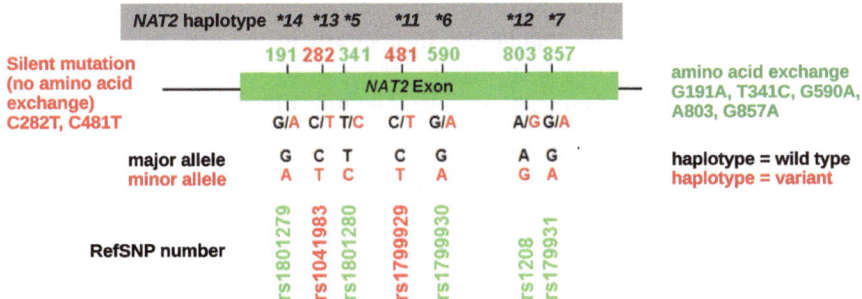

Fig. 6.3 Simplified structure of the NAT2 gene and their typical *single nucleotide polymorphisms (SNPs)*. RefSNP (Reference *SNP*) numbers identify single nucleotide polymorphisms in public databases. The notation G191A indicates that the common allele [G] was exchanged by the rare allele [A]. The number describes the corresponding nucleotide of the gene

the Europeans are unable to produce an important enzyme for the detoxification of a number of pollutants, including carcinogenic *polycyclic aromatic hydrocarbons (PAHs)*, and consequently have an increased risk for a number of diseases such as the development of urinary bladder cancer. *PAHs* are formed as a product of incomplete combustion of organic material, such as coal and fuel. So it is formed at large amounts in the production of coke.

Another gene that may play a role in the development of urinary bladder cancer is *N-acetyltransferase 2* (*NAT2*; cf. Fig. 6.3). The *NAT2* gene produces the *NAT2* enzyme, which, in turn, plays an important role in detoxifying chemicals that can cause urinary bladder cancer, in particular certain aromatic amines. The *NAT2* gene contains a number of *single nucleotide polymorphisms (SNPs)*, which, individually or in combination, cause the *NAT2* enzyme to differ in speed and efficiency of detoxification. In case of no genetic variants we refer to the rapid genotype, also called wild type (derived from plant and animal genetics). The slow genotype is present in about 50–60% of Europeans. In general, *NAT2* is one of the genes for which the effects of polymorphisms on metabolic performance can be easily and harmlessly determined in humans, in particular since the *NAT2* enzyme metabolizes not only potentially dangerous chemicals but also caffeine. Several hours after drinking coffee, a faster or slower metabolism can be inferred from the ratio of certain metabolites in the urine. One can even find metabolic differences between different slow genetic variants produced by different combinations of *SNPs*. In 2013, e.g., we found that a particularly slow genetic variant (*NAT2*6A*) leads to an increased risk of urinary bladder cancer even in low-pollution collectives when the metabolic capacity of the other slow variants of *NAT2* is already sufficient to detoxify the carcinogens.

'Genetic epidemiology' investigates causes of disease with a focus on genetic variants as risk factors. In this context, the variant of interest should be found less frequently in healthy persons than in persons affected by the disease. The greater the difference between these two frequencies, the greater the effect of the variant. If a disease is relatively common, we focus on so-called polymorphisms, i.e. variants with a frequency $\geq 5\%$ in population. In contrast, in the case of rare diseases or striking familial clustering in pedigrees, research focusses specifically on rare variants (frequency $< 5\%$ in the population). The following rules of thumb apply:

> The rarer a disease and the stronger the familial clustering, the rarer the variants and the stronger the influence on the probability of the disease (penetrance).
>
> The more frequent a disease and the lower the familial clustering, the more frequent the variants and the lower the influence on the probability of the disease.

In the case of bladder cancer, we are concerned with the second rule, i.e. we look for frequent genetic variants with each individual risk variant resulting in only a small risk increase. Large genome-wide studies suggest that for people carrying the genetic variant, the probability of the disease is twice as high as for people without the genetic variant.

If there are several polymorphisms of interest in the gene under investigation, as in *NAT2*, one often wants to find out which genetic variants are located together on one chromosome (cf. Fig. 6.2). In other words, we want to determine the haplotype from the available data, i.e. the pattern of genetic variants located on the maternal or paternal chromosomes. Often, this information cannot be obtained directly from the data. However, this problem can be solved in large studies for the most common genotypes using haplotype reconstruction algorithms. These algorithms start identifying *SNP* combinations that lead to unique haplotypes, which means that each *SNP* combination results in exactly one haplotype pair as solution. This information is then used to assign the most probable haplotype pairs to the non-unique *SNP* combinations, i.e., the combinations that allow for two or more haplotype pairs. In *NAT2*, 2/3 of the samples are non-unique, leaving 1/3 of the samples to be informative for the haplotype distribution.

6.3   Gene-Environment Interactions

Coming back to the main aim of the chapter, we focus on 'gene-environment interactions' in the remainder of the chapter. A gene-environment interaction is defined by the fact that the effect of the environmental factor changes, depending on the genotype, and vice versa. For example, the genetic factors *NAT2* and *GSTM1* are not independent influencing factors, but can interact with environmental influences. While genetic factors are easy to measure, this is much more difficult for environmental influences. In general, the exact exposure level and duration are much more difficult to determine for individual people than for groups. If variants in genes of xenobiotic metabolism such as *NAT2* or *GSTM1* are only relevant in interaction with a certain level of environmental exposure, it is possible to indirectly infer the intensity of such exposure. However, the latency period of a disease needs to be taken into account. In the case of cancer, latency periods of several years to decades may well apply, depending on the level of environmental exposure.

In studies in which the variant frequency is similar in cases and controls, it can be assumed that there was also no increased exposure during the corresponding period of disease development. Conversely, a higher frequency of the genetic variant in the diseased individuals compared to the control individuals would indicate an increased exposure of the population to the pollutant in question.

The amount of exposure may be occupational. In this case, those occupations which come with a high exposure to the pollutant and are particularly frequent in the patient group should be identified. More difficult is the situation for environmental pollution. In this case, exposure may be approximated by means of further characteristics such as age and place(s) of residence as well as the time of the first diagnosis. Such procedure may help in the planning and evaluation of further studies.

A recent study investigates the connection between urinary bladder cancer and structural change in Dortmund, a former center of the coal, iron, and steel industries. Until the 1990s, the area was strongly burdened both occupationally and environmentally. Accordingly, we were able to determine a significant increase of the *GSTM1* negative genotype in urinary bladder cancer cases from this area. This suggests that in a particularly large number of bladder cancer patients, an exposure to carcinogenic *GSTM1* substrates together with the lack of the enzyme needed for detoxification of these substances was causative for the development of cancer. In contrast, in our more recent studies conducted after the demise of the coal, iron, and steel industries, the variants of the *GSTM1* gene have the same frequency in cases and controls (Fig. 6.4).

The starting point was an older study conducted in Dortmund between 1992 and 1995, which revealed an exceptionally high proportion of 70% negative *GSTM1* genotypes in urinary bladder cancer cases, in contrast to 54% in the control group. This corresponds to an odds ratio (OR) of 1.99, i.e. a doubling of the risk of urinary bladder cancer in the affected patients. For the two more recent studies in Dortmund (2009–2010 and 2012–2013), the question was, therefore, whether this high proportion could be related to the high exposure to the pollutant at the workplace or in the environment. Given the usual latency period of urinary bladder

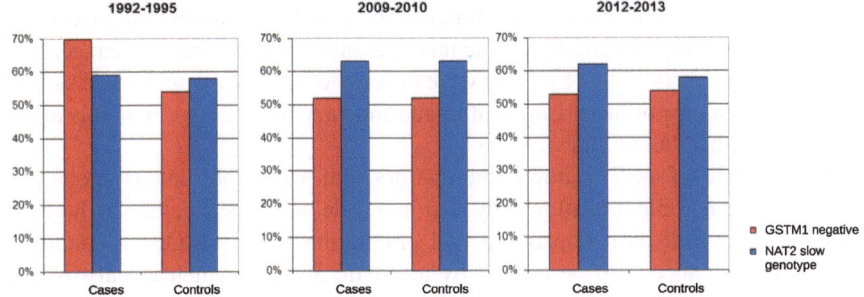

Fig. 6.4 Structural change in Dortmund: results of three urinary bladder cancer studies; red = *GSTM1* negative, blue = *NAT2* slow genotype

cancer, the proportion of *GSTM1* negative genotypes should be similar to that of controls 15–25 years after the structural change.

In the two more recent Dortmund studies, *GSTM1* negative genotypes were alike in cases and controls and corresponded to portions of 50–55%, as can be found in European populations in general. The 2009–2010 study showed 52% in cases and 52% in controls, the 2012–2013 study 53% in cases and 54% in controls (cf. Fig. 6.4). This corresponds to odds ratios of 0.99 and 0.95, respectively. These findings coincide with the previous, no longer existing, back then increased exposure to *PAHs* due to emissions from the coal, iron, and steel industries. Furthermore, the slow and fast genotypes of the *NAT2* gene were also investigated (cf. Fig. 6.4). Here, no differences between cases and controls could be found in the older study (1992–1995: NAT2 slow: 59% cases, 58% controls), which indicates that the carcinogenic substrates of *NAT2* did not play a major role in the period of investigation in general but in some individual cases only, e.g. in heavy smokers or in participants who suffered high or long-term occupational exposure. In the most recent study from 2012 to 2013, we observed a higher proportion of *NAT2* slow genotypes among the urinary bladder cancer cases of 62% when compared to 58% among the controls, but this is completely within the usual range of variation. This is also confirmed by an odds ratio of 1.18 and the non-significant p-value of the chi-square test, with $p = 0.4217$. Thus, the structural change has not had an effect regarding the substrates of *NAT2*.

6.4   Conclusion

The tendency towards high proportions of the negative *GSTM1* genotype as a marker for high exposure to *PAHs* is not only evident in Dortmund but also in other highly industrialized regions and cities with high air pollution. For example, a study on the highly industrialized area around Brescia, Italy, between 1997 and 2000 also reports a very high proportion of *GSTM1* negative genotypes among urinary bladder cancer cases (66%), in particular when compared to a control group (52%).

Regarding *NAT2*, in the Los Angeles Bladder Cancer Study (initial diagnosis period: 1987–1996) the situation is similar to the *GSTM1* situation in the Dortmund study. In the Los Angeles study, significant differences between bladder cancer cases (58%) and controls (49%, OR = 1.58) were described, too. Los Angeles has suffered from significant air pollution from traffic and industry for decades, which is among the highest in the United States.

The example of *NAT2* shows that factors that were relevant as modulators for decades may suddenly play only a minor role with the decline of high-risk occupations as well as occupational and environmental exposures. As indicated above, bladder cancer can be caused by aromatic amines, which are also part of tobacco smoke. In this respect, for decades smokers with a slow genotype had an additional risk of urinary bladder cancer compared to smokers with a fast *NAT2* metabolism. Meanwhile, a lower influence of *NAT2* is observed in this context as well, which can only be detected by the more sensitive ultra slow variant of *NAT2*.

In this respect, differentiation into the various haplotypes of *NAT2* has become increasingly important, even if exposure to aromatic amines is decreasing. In a constantly changing environment, other risk factors become more and more important. For years, physicians and epidemiologists have observed an increase in an almost extinct disease in Europe: tuberculosis is on the rise again. The relationship to *NAT2* is the following: *NAT2* metabolizes many substances, e.g. important antibiotics. If these are metabolized too slowly by the slow *NAT2* variants, this results in a risk of irreversible liver damage, as is the case with the indispensable antituberculosis drug *isoniazid*.

6.5 Further Reading

The research in this chapter was carried out jointly at the Leibniz Research Centre for Working Environment and Human Factors at TU Dortmund University (IfADo) and the Faculty of Statistics at TU Dortmund University. Our special thanks for helpful discussions go to Jan Hengstler, head of the Toxicology/Systems Toxicology research area at IfADo. Details on the results can be found in Golka, Reckwitz, Kempkes, Cascorbi, Blaskewicz, Reich, Roots, Sökeland, Schulze, Bolt (1997): "N-acetyltransferase 2 (NAT2) and glutathione S-transferase μ (GSTM1) in bladder-cancer patients 5in a highly industrialised area" in International Journal of Occupational and Environmental Health 3, 105–110, for the 1992–1995 study in Ovsiannikov, Selinski, Lehmann, Blaszkewicz, Moormann, Haenel, Hengstler, Golka (2012): "Polymorphic enzymes, urinary bladder cancer risk, and structural change in the local industry" in Journal of Toxicology and Environmental Health Part A 75(8–10), 557–565, for the 2009–2010 study, and in Krech, Selinski, Blaszkewicz, Bürger, Kadhum, Hengstler, Truss, Golka (2017): "Urinary bladder cancer risk factors in an area of former coal, iron, and steel industries in Germany" in Journal of Toxicology and Environmental Health Part A 80(7–8), 430–438, for the 2012–2013 study. The Brescia study is reported in Hung, Boffetta, Brennan, Malaveille, Hautefeuille, Donato, Gelatti, Spaliviero, Placidi, Carta, Scotto di Carlo and Porru (2004): "GST, NAT, SULT1A1, CYP1B1 genetic polymorphisms, interactions with environmental exposures and bladder cancer risk in a high-risk population" in International Journal of Cancer 110(4), 598–604. The results from Los Angeles are described in Yuan, Chan, Coetzee, Castelao, Watson, Bell, Wang, Yu (2008): "Genetic determinants in the metabolism of bladder carcinogens in relation to risk of bladder cancer" in Carcinogenesis 29(7), 1386–1393.

Chapter 7
Statistics and the Maximum Human Lifespan

Jan Feifel and Markus Pauly

Abstract Is there a natural limit to the maximum human age or can we theoretically live forever? Understanding and modeling human mortality is of great social and economic interest. In the following chapter, we explain how to estimate the maximum human lifespan using statistical methods and provide possible answers to the two introductory questions.

7.1 Background

Often in scientific studies, we are not only interested in the most frequent or average values of a feature under observation but also wants to quantify possible maximum or minimum values. Indeed, we frequently encounter extreme events in our daily lives. For example, a glance at the news headlines promises a "new negative record in debt," the "lowest economic growth in 44 years," "the hottest summer ever," or "the heaviest precipitation on record." Some of these sensational headlines reporting extreme events may only be geared towards increasing sales numbers of newspapers, etc. Still, the question arises whether future events could be even more extreme than they currently are. For example, under increasingly extreme weather conditions, people living in coastal regions might ask how high a new dam must be built to reliably keep out future floods.

Such questions can be addressed by means of statistical extreme value theory. This subfield of statistics received a lot of attention in the tabloid press at the beginning of the twenty-first century, after two well-known Tilburgian statisticians had estimated the potential extreme values for 21 world records of track and field

J. Feifel (✉)
Ulm, Germany
e-mail: jan.feifel@alumni.uni-ulm.de

M. Pauly
TU Dortmund, Department of Statistics, Dortmund, Germany
e-mail: markus.pauly@tu-dortmund.de

athletics. Knowing these findings, one could brag about the fact that, e.g., men will probably never run faster than 9.29 s in the 100 m and women will not throw the discus further than 85 m. One could thus discuss how outstanding the current world records really are, when, e.g., talking to colleagues during the coffee break. A special challenge for the statistical analysis of extreme values is the sparse information on the extremes of events. These, by nature, occur rather rarely, as, e.g., world records or extreme weather conditions such as floods and draughts. Moreover, statisticians want to predict extreme developments above or below the observation range, which is far more challenging than simple mean estimations.

Such statistical challenges become even more important when aiming to determine, e.g., the maximum age of current or future generations of humans. This does not only concern physicians and demographers, but also insurance companies. The latter want to protect themselves by estimating how long they have to pay insurance benefits for their clients. In this context, a nice anecdote exists about André-François Raffray, who thought of himself as a smart businessman when he concluded a contract with the 90-year-old Jeanne Calment, in Arles, France, in the 1960s. She received 2500 francs each month in return for her apartment, where she was allowed to live until her death. However, when Raffray died 30 years later, the old lady was still alive and he had already paid 920,000 francs for an apartment in which he never lived. Mrs. Calment died in 1997 at the age of 122 years and 164 days and, as of today, holds the official record of maximum human age.

Such a comparatively early und so far unchallenged record is astonishing since one often reads that new medical developments steadily improve life expectancy. However, such improvement mostly refers to the average life expectancy, which, unfortunately, has little significance for determining the maximum age. Therefore, it is unclear whether an increase can also be observed for the maximum age and whether one could theoretically live forever. From a statistical point of view, the methods of extreme value theory offer an attractive approach to investigate these questions in more detail.

7.2   From the Average to Extreme Value Theory

As is always the case in applied statistics, such questions are answered drawing on available data. Extreme values of past events can be used as reference level. However, we ultimately want to estimate extremes that are not included in the existing data. The mean value, which is possibly the most often reported data characteristic, does not provide the desired insight. The maximum value of a data sample comes much closer to our goal. However, it is still not sufficient for predicting future extremes, as the observed maximum does not necessarily correspond to the possible extreme of a distribution.

Let us consider the following simple example: We are interested in the tallest student of a school. To collect our sample, we measure the height of every student who enters the school building. However, the tallest student may not have been

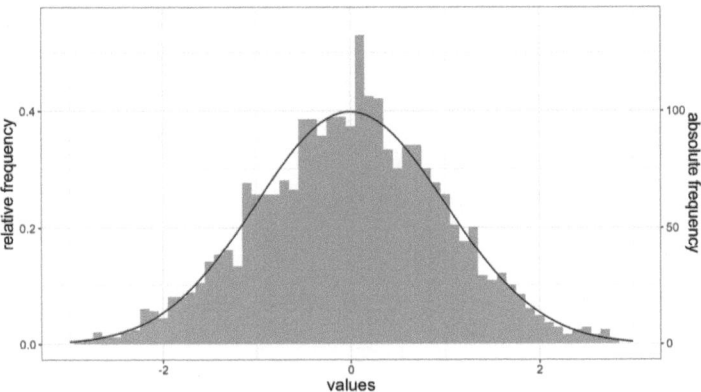

Fig. 7.1   1000 mean values from 1000 random numbers each, plotted together with an approximate Gaussian distribution

among those if, e.g., he/she was sick on the day of the measurement. Nevertheless, by means of statistical methods, we can estimate relatively accurately the body size of the tallest student on the basis of the sizes of the observed students.

To further illustrate this point, we would like to simulate random numbers from a given distribution of values and determine their corresponding extreme values on a computer. The most well-known distribution is the Gaussian distribution (also known as the normal distribution), which can be described by a bell-shaped curve. This so-called density curve can be understood as an approximation of the distribution of the sample mean (see Fig. 7.1). In many cases, this is very useful since the true distribution of the data and the corresponding mean are usually unknown. Roughly speaking, the density of a distribution indicates how likely it is to observe a certain value.

In a similar way, approximations can be derived for the distributions of maxima. In this case, the situation is more complicated than for the distribution of the mean, since for extreme values different types of distributions can occur, depending on what we observe.

One of these distributions is the so-called Gumbel distribution, which is used, e.g., in hydrology to model the maximum precipitation at a certain location. By means of this distribution, we want to illustrate the extreme value problem. In Fig. 7.2, we have artificially generated random numbers (points on the x-axis) which are generated by means of the Gumbel distribution. These were plotted against the density (distribution curve), which shows the frequency of the distribution relative to the value on the x-axis. Here, the largest simulated random number is approximately 67. However, if one looks at the enlarged part of the graph, which starts at a value of 55, it becomes apparent that the density continues towards zero. This indicates that larger values, e.g. 75, would have been possible in a different sample. Thus, the largest observed value alone is not sufficient to describe the maximum possible

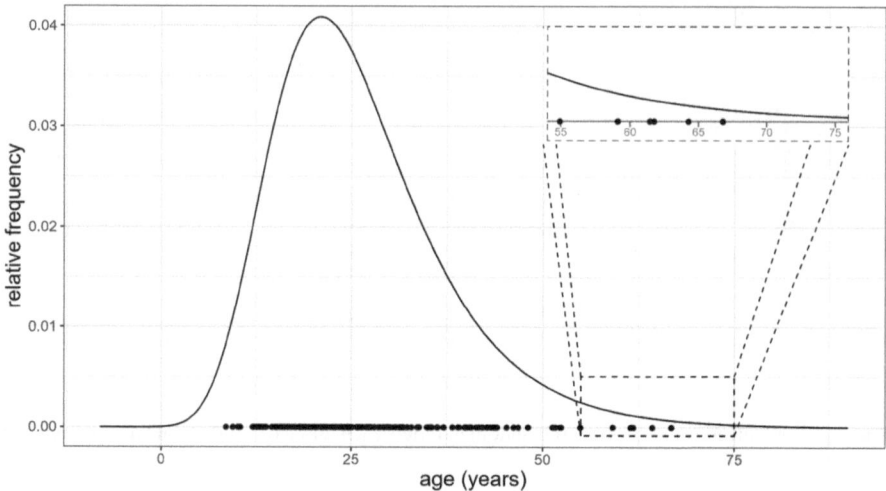

Fig. 7.2 Density of a Gumbel distribution and computer-generated random numbers (points) from this distribution. The part from 55 to 75 is highlighted and enlarged

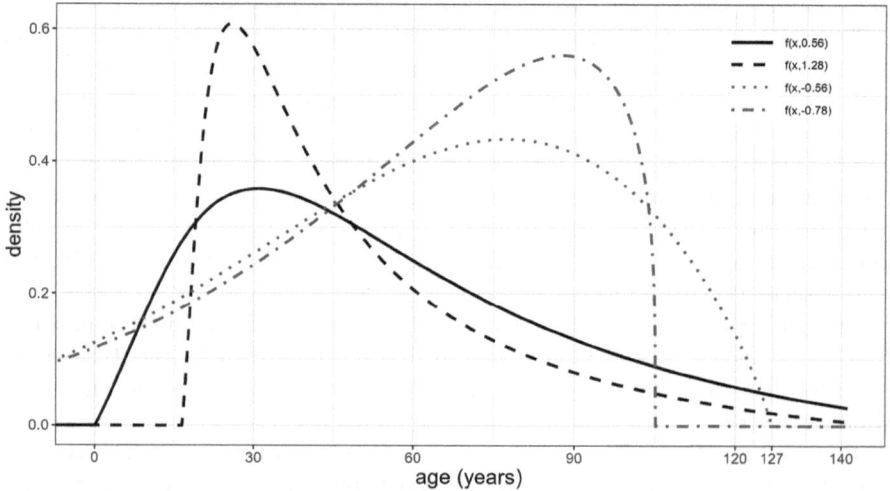

Fig. 7.3 Different densities f(x,γ) of the (shifted and scaled) generalized extreme value distribution with positive (black curves) and negative (gray curves) extreme value index γ

value of a distribution. However, it is still considered valuable for the estimation of extreme values.

Using mathematical methods, it can be shown that the extrema of an arbitrary distribution can be approximated by the density of a generalized extreme value distribution for a sufficiently large sample. The behavior of this density mainly depends on a single value γ, the so-called extreme value index. For the case of maximum human age, Fig. 7.3 shows realizations of different densities of extreme value distributions. The sign of the extreme value index indicates whether the distribution theoretically allows for increasing values or whether there is a maximum endpoint: A positive extreme value index ($\gamma > 0$) represents a distribution density that permanently lies above the x-axis so that arbitrarily large values can be observed. If the extreme value index is negative ($\gamma < 0$), there is an endpoint. Figure 7.3 shows the density for different values of γ. The densities are shifted and scaled to be of importance to our question. The question arises how to determine the unknown extreme value index. For this purpose, statisticians have to find a good approximation (estimate) to this unknown index, based on all available data. In a second step, a potential maximum value will be estimated. There are many ways to estimate γ, which differ, e.g., in the number of largest observations included in the approximation.

After this rather theoretical excursion, the question arises what this means for estimating the maximum life expectancy of human beings. We can use the above approach to investigate whether we could theoretically live forever ($\gamma > 0$) or whether our life is limited by a maximum age ($\gamma < 0$). In the second case ($\gamma < 0$), we would be able to estimate the exact maximum age by taking into consideration existing age records.

Despite all efforts of modern medicine, infinite life seems rather impossible for humans. However, 100 years ago, a maximum age of 120 seemed similarly impossible. To investigate maximum life expectancy in more detail, it is unreliable to combine demographic death statistics from all over the world, since sources are of different quality and thus reliability. Moreover, death statistics from around the globe would be far too heterogeneous due to different living standards and life expectancies. For example, inhabitants of the African continent are considered to have a lower life expectancy than Europeans or North Americans. Therefore, we analyzed mortality data for female U.S. citizens from a specific time period to obtain a large enough, homogenous data sample. We will present the results at the end of this chapter and first explain some additional challenges we faced in our analysis, when compared to the theoretical considerations presented above.

7.3 Challenges in Working with Demographic Data

The quality of data needed to investigate the life expectancy of humans is questionable. Due to data privacy reasons, data on death dates are strongly protected and usually provided in anonymized form by authorities only. Furthermore, most of the female U.S. citizens in our study who passed the age of 100 were born in the eighteenth century. At that time, accurate documentation of birthdays was not very common and home births occurred frequently, which further complicated an official recording of early life data. First officially validated results were often provided by a census, in which individuals were listed a few years after birth or in which their age was estimated.

To ensure good data quality, we, therefore, restricted our study to mortality data from 1980 onwards, based on two different acknowledged databases: the International Database on Longevity (IDL) and the Human Mortality Database (HMD). In the IDL, data of so-called supercentenarians were collected until 2003. Supercentenarians can be considered human dinosaurs, i.e. people who have reached an age of at least 110 years. For the period from 1980 to 2003, the IDL includes death data of 309 U.S. women. The records contain the exact, officially verified birth and death dates and thus guarantee very high data quality. However, for our purposes and in particular for answering the question whether human life expectancy has risen over the last decades, the number of data points is far from sufficient. For this reason, we consulted the HMD as an additional database. It contains almost 26 million death records of U.S. women for the period from 1980 to 2003. In contrast to the IDL, however, the data structure and quality are quite different. In particular, people older than 110 years are censored and listed as 110+, which is rather vague for the purpose of our study. For our data set, this concerns 1853 people. Next to those people for whom no exact death date was indicated, various others were still alive but older than 110. Such people also influence the estimation of the maximum age and have to be taken into account in the statistical procedure. This can be explained very simply: Let us imagine that Jane Doe was still alive at the age of 120 but that the earlier maximum age was 119. The lifetime of Mrs. Doe would then set an age record at death and would therefore be relevant for estimating the maximum age. Modern statistical methods can incorporate such findings into the estimation.

7.4 Results

Applying these methods to the combined data set of HMD and IDL data of U.S. females (cf. Sect. 7.3), we obtain the following results: First, we can rule out an infinitely long life from a statistical point of view, i.e., it is highly unlikely. This finding is consistent with a large number of earlier demographic studies, but it contradicts others. However, our study is the first to provide a sound statistical justification for such a claim using extreme value theory. As a maximum attainable age, we were able to identify an estimate of 127 years. Taking into consideration the natural uncertainty of the estimate, a maximum age of 131 years seems not entirely implausible to us. This is quite remarkable, considering that the oldest American woman so far lived for 119 years. Furthermore, we surprisingly did not find a rise in maximum age over time for the time span from 1980 up to 2003.

7.5 Conclusion

Statistical methods enable us to make statements about values which we have not observed, since they lie beyond the observed sample. They are nevertheless—or precisely because of this particular property—relevant and interesting for extreme value estimation. We illustrated the methods on the basis of our study on maximum life expectancy and could thus answer the question "Can we live forever?" with a (significant) 'No'. Those who find this answer too pessimistic should reflect that the hope for eternal life on earth is currently not so tempting: This is not only addressed in the song "Who wants to live forever?" by the British band Queen but also in current scientific literature, which addresses issues of low quality of life in old age.

7.6   Further Reading

Advanced statistical methods are becoming increasingly popular in demography. Our results are presented in detail in the preprint of J. Feifel, M. Genz, and M. Pauly (2018): "The myth of immortality: An analysis of the maximum lifespan of US Females," ifa Ulm, University of Ulm. It also gives an overview of other estimated ages. For the sporting achievements mentioned in Sect. 7.1, see J. Einmahl, and J. Magnus (2008): "Records in athletics through extreme-value theory," Journal of the American Statistical Association 103:484, 1382–1391. For the mathematically interested reader, we recommend a look at the book by L. De Haan and A. Ferreira (2007): "Extreme value theory: an introduction," Springer Science & Business Media, e.g., to understand the shape of the extreme value distribution or to look up details on different estimates of the extreme value index. The apartment anecdote about Jeanne Calment (see Sect. 7.1) we have taken from the 1995 New York Times newspaper article "A 120-Year Lease on Life Outlasts Apartment Heir."

Part II
Sports and Entertainment

Chapter 8
Statistics and Soccer

Andreas Groll and Gunther Schauberger

Abstract One of the reasons why people are fascinated by soccer is the fact that chance plays a greater role here than in many other sports. Despite this (or maybe because of it?), soccer is also an ideal object for statistical investigation.

8.1 More Goals by Means of Statistics

In 2014, the Danish soccer club FC Midtjylland changed its club policy. Since then, the club's management has made all major decisions on player transfers and tactical strategies based on statistical models. After near-bankruptcy in 2012, this approach was rewarded with the Danish championship and the qualification for the UEFA Champions League. This is an impressive example of how interest in scientific analysis of soccer data has increased in recent years. Some of these data are collected by the clubs themselves, for example by taking medical measurements of players during training or by using video and other technology systems to capture relevant player movements. Other data are compiled by companies such as Opta, Amisco, etc., from which the clubs buy them.

Coaches and clubs hope to gain new insights from these data and improve their teams' chances of success. In addition to pure numbers and tables, more and more complex and sophisticated mathematical models are being used. In the present chapter, we present some findings on match results and team-related 'macro variables'.

A. Groll (✉)
TU Dortmund, Department of Statistics, Dortmund, Germany
e-mail: groll@statistik.tu-dortmund.de

G. Schauberger
TU Munich, Epidemiology, Munich, Germany
e-mail: gunther.schauberger@tum.de

C. Weihs et al. (eds.), *Statistics Today*, Society, Environment and Statistics,
https://doi.org/10.1007/978-3-662-68907-3_8

8.2   A Statistical Model for Predicting Goals

The main aim in soccer is scoring goals. Figure 8.1 shows the relative frequencies of home (red) and away (blue) goals for four Bundesliga seasons from 2014/2015 to 2017/2018. Unsurprisingly, home teams score more goals than away teams: the frequencies of scores of two or more goals is higher for home teams (red) than for away teams (blue).

To model such a random number of goals, the Poisson distribution is a reasonable choice. It comes with the probabilities

$$\mathbb{P}(X = n) = \frac{\lambda^n e^{-\lambda}}{n!}, \quad n = 0, 1, 2, 3, \dots, \tag{8.1}$$

where X denotes the number of goals. The distribution therefore indicates the probability that exactly zero, one, two, three, or however many goals are scored in a game. The parameter λ reflects the average goal rate of a match and determines whether high numbers of goals have a high or low probability. For the four Bundesliga seasons 2014/2015–2017/2018, the values $\lambda_H = 1.60$ and $\lambda_A = 1.23$ represent the average number of home and away goals, respectively. Hence, for

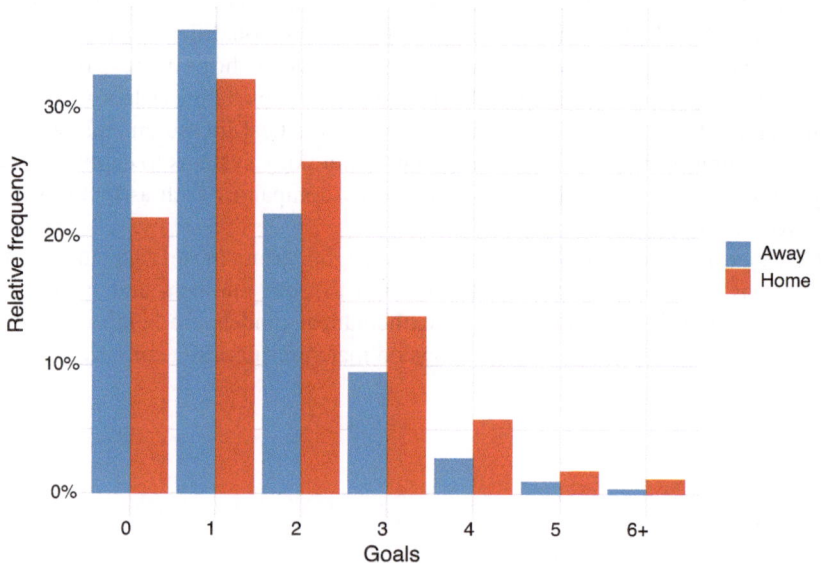

Fig. 8.1 Relative frequencies of home and away goals for the four Bundesliga seasons 2014/2015–2017/2018

example, the probability that the home team will score exactly two goals in a randomly selected Bundesliga match is

$$\mathbb{P}(X_H = 2) = \frac{(\lambda_H)^2 e^{-\lambda_H}}{2!} = \frac{1.6^2 e^{-1.6}}{2} \approx 0.26 = 26\% \,.$$

In Fig. 8.2, in addition to the observed relative frequencies of some selected match results (dark), their probabilities according to the underlying Poisson distributions (light) are shown. One can see that in our example the probabilities according to the two Poisson distributions match the observed relative frequencies quite well. Thus, based on the average goal rates alone, it is already possible to predict quite accurately how many home and away goals will be scored in total.

The next step is to focus on specific match results and calculate, for example, the probability of a 0:1 win in a randomly selected Bundesliga match. In the simplest case, the two corresponding probabilities are multiplied, here 20% for 'no home goal' and 36% for 'exactly one away goal' scored by the opposing team, i.e. 0.2·0.36 = 0.072. However, this only works if we assume the two events to be independent.

8.3   Influential Variables

The website http://www.kicker.de/ of the German soccer magazine *kicker* provides a summary of numerous match characteristics right after each Bundesliga match. Table 8.1 provides an example of some standard characteristics.

For the match Borussia Dortmund against Bayern München on the 11th match day of the Bundesliga season 2016/2017, it is immediately apparent that Borussia Dortmund was inferior in almost all respects, but still won the match by a close shave (Table 8.1). In particular, Bayern München was clearly superior in terms of completion rate, possession, tackling rate, and shots on goal, which are criteria often discussed among soccer fans as indicators of success. Only with respect to running performance, Dortmund had a slight edge. So, is this game to be considered an outlier? Or is it possible that the running performance is particularly important? And what role do the other criteria play that are so often mentioned and noted among soccer fans and sports commentators, such as shots on goal, completion rate, and ball possession?

Let's take a look at the criterion 'running performance'. At first glance, one would assume that teams that run more and thus put in a greater effort also win games more often on average. Figure 8.3 shows the running performance over the 34 match days of the 2017/2018 season of Bayern München (champion at the end of the season), Borussia Dortmund (ranked 4th), SV Werder Bremen (ranked 11th), and Hamburger SV (ranked 17th). Additionally, for each match day, the outcome of the match is indicated by the different symbols (circle, triangle, and square), from the respective club's perspective.

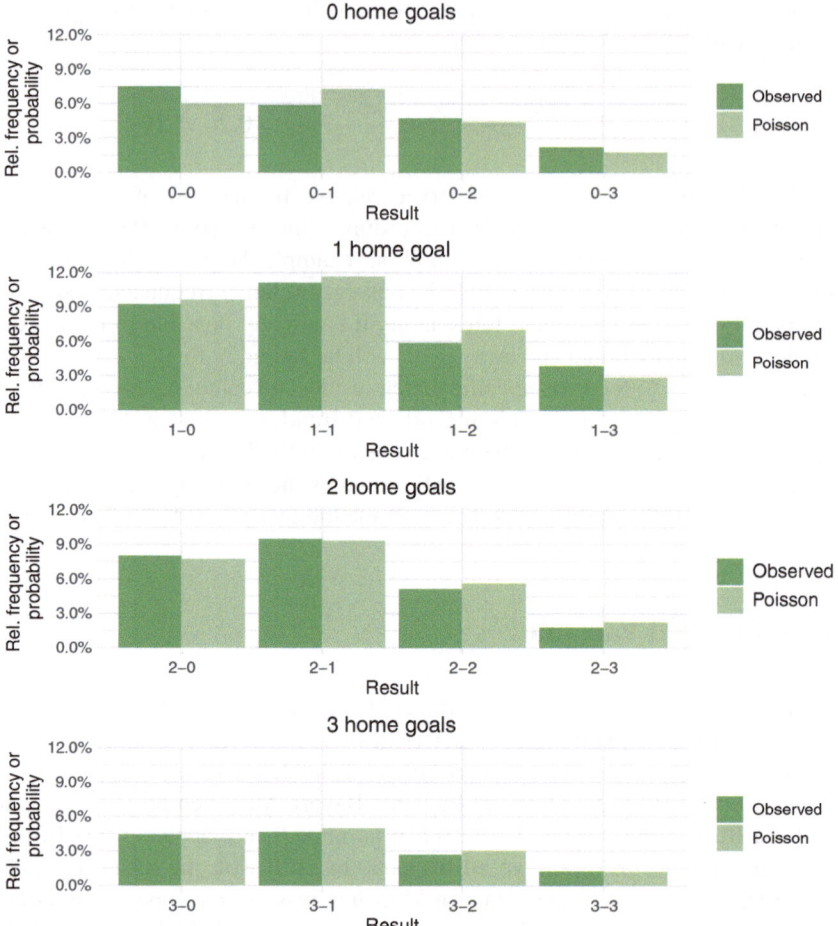

Fig. 8.2 Relative frequencies of 16 selected match results for the four Bundesliga seasons 2014/2015–2017/2018 (dark) compared to the goal distribution according to two independent Poisson distributions (light)

First, it is noticeable that the two more successful teams, Bayern München and Borussia Dortmund, were relatively stable in their running performance at the beginning of the season. However, from around match day 10 onwards, substantially larger fluctuations occurred. While the number of matches won by Bayern München significantly increased and stabilized, the opposite was true for their running performances. Those only stabilized again over the winter months.

A very similar picture emerges for Borussia Dortmund. After a highly variable phase between match day 9 and 15 in the winter months, the running performance first also settled at relatively high values, before decreasing again towards the end of the season and becoming substantially more variable again. For both clubs,

Table 8.1 Match characteristics for the Bundesliga match Borussia Dortmund vs. FC Bayern München from match day 11 of the Bundesliga season 2016/2017 (From: http://www.kicker.de/, logos reprinted with kind permission by ©BVB 09 Dortmund and ©FC Bayern München [2019]. All Rights Reserved))

Goals	**1** : 0	Goals
Shots on goal	11 : **18**	Shots on goal
Running performance	**116.95** : 115.62	Running performance
Played passes	348 : **695**	Played passes
Completed passes	245 : **592**	Completed passes
Misplaced passes	103 : 103	Misplaced passes
Completion rate	70% : **85%**	Completion rate
Possession	34% : **66%**	Possession
Tackling rate	41% : **59%**	Tackling rate
Foul/Hand played	14 : 14	Foul/Hand played
Fouls received	13 : 13	Fouls received
Offside	0 : **4**	Offside
Corner kicks	3 : **9**	Corner kicks

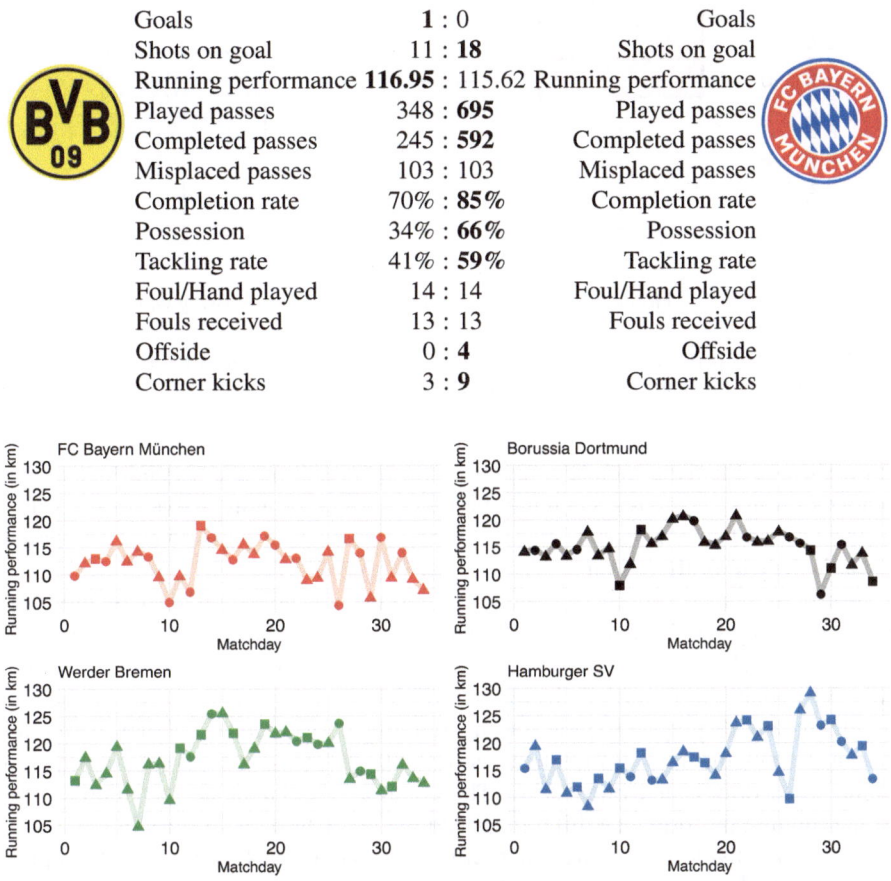

Fig. 8.3 Running performance over the 34 match days of the 2017/2018 season of four selected Bundesliga teams; ● = win, ▲ = draw, ■ = defeat

extreme values of both very high and, less frequently, very low running performance only occur in victories or defeats, whereas draws come along with more moderate running performances. Thus, a correlation between high running performance and success is not discernible at first glance. Even in comparison to SV Werder Bremen and Hamburger SV, who performed significantly worse at the end of the season, Borussia Dortmund's and Bayern München's players do not seem to run noticeably more, but rather less.

For Bremen and Hamburg, particularly strong running performances also occurred in games that ended in a draw. While Werder Bremen, after a variable

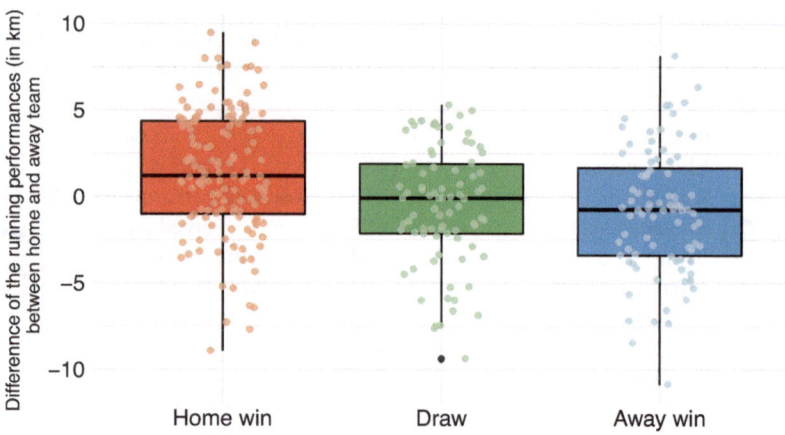

Fig. 8.4 Running performance differences between home and away teams for different outcomes for all matches of the four Bundesliga seasons 2014/2015–2017/2018

start, showed relatively stable running performances from the second half of the season onwards, Hamburg's running performances were quite stable, in particular in the first half of the season. For all teams, the running performance decreased towards the end of the season.

While individual progress curves (Fig. 8.3) do not indicate a clear relationship between running performance and ultimate success, a comparison of the differences in running performance between the home and away teams and the match results (Fig. 8.4) creates a somewhat clearer picture.

Figure 8.4 displays boxplots of the running performance differences for all matches from the four Bundesliga seasons 2014/2015–2017/2018, according to home wins, draws, and away wins. The colored boxes indicate the middle 50% of the running performance differences and the black horizontal lines in the box—the median—divides the entire diagram into two areas, each illustrating the upper and lower 50% of the running performance differences. The vertical lines—the whiskers—outside the box show the range where the data do not belong to the middle 50% but do not yet constitute extreme outliers. Running performance differences that lie above or below these whiskers represent the extreme outliers of the observed data and are shown as black dots. For example, Fig. 8.4 illustrates such an outlier for a game that ended in a draw and in which the away team ran substantially more than the home team. The colored dots in Fig. 8.4 additionally indicate the actually observed, individual differences in running performance.

Obviously, the home team's running performance exceeds the performance of the away team most distinctly for home victories, since the clearly larger proportion of the observed differences lie above 0. For draws, on the other hand, the box almost symmetrically ranges around the zero line, while for away victories, the higher running performance was achieved by the away team. Thus, a high running performance and a win often go hand in hand.

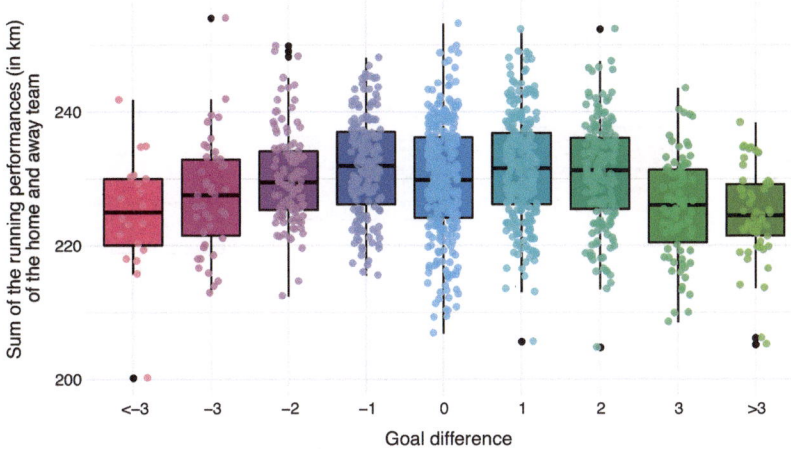

Fig. 8.5 Sum of the running performances of both teams at different goal differences (goals home team minus goals away team) for all matches of the four Bundesliga seasons 2014/2015–2017/2018

However, if we take a closer look, the relationship between running performance and success turns out to be more complex. If we plot the sum of the running performances of both teams in a match against the difference in goals scored by the home and away teams (see Fig. 8.5), we see that the sum of the running performance of both teams clearly decreases for both distinct away wins (the two leftmost boxplots) and distinct home wins (the two rightmost boxplots). This can presumably be explained by the fact that a team leading by at least three goals may go into defense and reduce its attacks, while the opposing team usually surrenders to its fate demoralized. Accordingly, the running rates of both teams are particularly high for games ending in a narrow lead. Draws (yellow box in the middle) tend to be associated with a somewhat lower running performance again, since it may happen that at the end of the game both teams are satisfied with a draw and no longer willing to take unnecessary risks. This shows that the causality between running performance and outcome of a match is not always straightforward.

Thus, the question arises whether there might be other variables besides the running performance that show a clearer connection to success and that are able to better differentiate, e.g. between home wins, draws, or away wins. In analogy to our discussion above (Fig. 8.5), we consider a comparison of the differences in shots on goal between home and away teams and the match result. As can be seen in Fig. 8.6, the result is somewhat surprising. It shows that in games that ended with a home win, the home team also shot on goal more often on average than in games that ended in a draw or an away win. However, even in games that ended in a draw, the home team shot on goal more often. In games that ended in an away win, the away team only had more shots on goal in about half of the cases. In the two latter scenarios, it thus seems that the home team basically creates more shots on goal, without necessarily being able to convert them into goals.

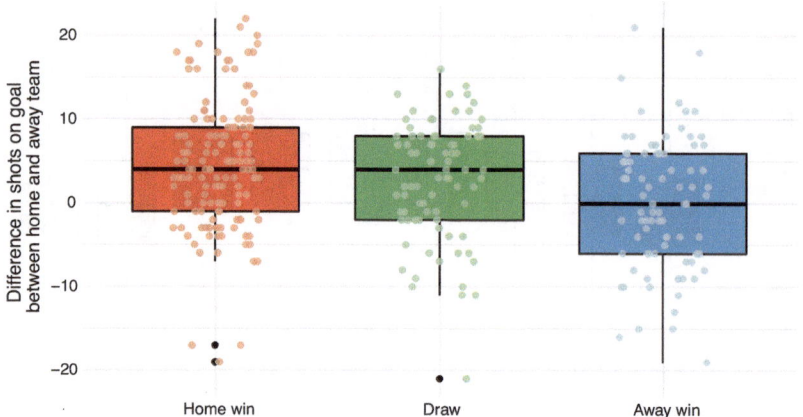

Fig. 8.6 Difference in shots on goal between home and away teams for different match results for all matches of the four Bundesliga seasons 2014/2015–2017/2018

8.4 Conclusion

In general, the outcome of a soccer match is rather uncertain and this is exactly what makes it so fascinating. It is the interplay of many small random events, but in particular individual goals, sometimes scored rather randomly or simply by chance, and the psychology of individual players or the whole team, which ultimately determine the outcome of a game. Statistical methods may help unveil potential relationships between match characteristics and the outcome of matches, but despite sophisticated models and advances in the scientific analysis of soccer data, fans can still look forward to exciting games with surprising results in the future.

8.5 Further Reading

Section 8.2 has been inspired by the book "The Numbers Game – Why Everything You Know about Soccer Is Wrong" (Penguin Books, 2013) by C. Anderson and D. Sally, where many more exciting aspects and examples on the topic of soccer analytics can be found in an easy-to-understand language.

Chapter 9
The Players' Anxiety at the Penalty Kick: Who Is the Best Penalty Taker, Who the Best Goalkeeper?

Peter Gnändinger, Leo N. Geppert, and Katja Ickstadt

Abstract Penalty kicks create exciting moments in soccer, often resulting in goals. With the help of statistical models, we investigate which goalkeepers and which penalty takers are particularly successful in the German soccer Bundesliga and which other factors have an impact on the outcome of penalty shots.

9.1 Penalties in Soccer

Goalkeeper against penalty taker. It is a simple duel, but it can decide whole matches. A penalty can give birth to tragic figures—or to heroes. In Germany, Andreas Brehme is still well-known for scoring the only goal of the 1990 World Cup's final via penalty in the 85th minute. More recently, Harry Kane managed to score a penalty both in regular playing time and in the penalty shootout in England's win in the round of 16 match of the 2018 World Cup. Naturally, coaches like having players who excel in scoring penalties as well as goalkeepers who excel in saving penalties in their squad. But how can we determine which players show particular abilities in this regard?

The easiest way is to look at the relative frequency of converted and saved penalties, respectively. However, these numbers are not always meaningful. Frequencies do not necessarily say a lot about the underlying abilities, in particular when players were not involved in many penalties. Figure 9.1 illustrates how many goalkeepers and penalty takers were involved in how many penalties, based on all penalties in the German Bundesliga from its first season in 1963/1964 up until its 2016/2017 season. On the left-hand side, we can see that the large majority of goalkeepers had only few penalties taken against them. Such an imbalance is even more obvious for the penalty takers, shown on the right-hand side of the figure. While there is only

P. Gnändinger · L. N. Geppert · K. Ickstadt (✉)
TU Dortmund, Department of Statistics, Dortmund, Germany
e-mail: peter.gnaendinger@tu-dortmund.de; geppert@statistik.tu-dortmund.de;
ickstadt@statistik.tu-dortmund.de

C. Weihs et al. (eds.), *Statistics Today*, Society, Environment and Statistics,
https://doi.org/10.1007/978-3-662-68907-3_9

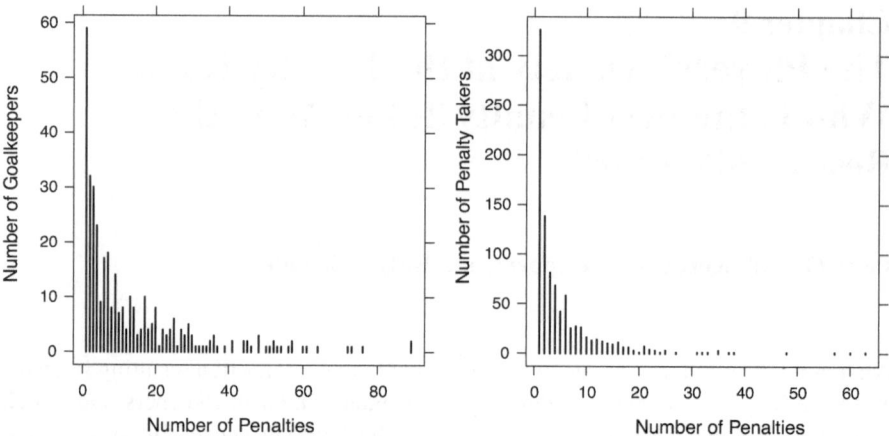

Fig. 9.1 Number of penalties per goalkeeper (left-hand side) and per penalty taker (right-hand side) for all penalties in the German Bundesliga from 1963/1964 to 2016/2017

one goalkeeper, every player currently on the pitch is allowed to shoot a penalty. The number of penalty takers is thus a lot higher. This is underlined by the observation that more than 300 players took just one penalty each in the investigated time span.

When employing relative frequencies, goalkeepers are rated best when they had only one or two penalties taken against them and saved them all. On the other hand, goalkeepers who allowed a goal from the only penalty against them exhibit the lowest relative frequency possible.

An alternative to relative frequencies are statistical regression models. These models allow ranking goalkeepers as well as analyzing the influence of additional variables on the estimation of goalkeepers' abilities of saving penalties and can also take into account takers' abilities of converting penalties. We will describe two suitable models in more detail in Sects. 9.3 and 9.4. In Sect. 9.3, we will describe Model 1 for the analysis of variables associated with the goalkeepers' saving abilities. This model is suitable for ranking goalkeepers but not for ranking penalty takers. For that reason, we will study the more complex Model 2 for ranking both goalkeepers and penalty takers in Sect. 9.4.

9.2 The Penalty Data Set

We obtain the necessary data for our regression models from the websites of the German sports magazine kicker (www.kicker.de) as well as transfermarkt (www.transfermarkt.de). Since the beginning of the German Bundesliga until the end of the season 2016/2017, a total of 4599 penalties was taken, of which 3432 were converted and 856 were saved by the goalkeeper. The remaining 311 penalties hit the post or missed the goal. We will exclude these from our analysis in the following. A total of 347 goalkeepers and 871 penalty takers were involved in at least one penalty.

In addition to the teams and the names of the players involved, we know the season, the match day, and the minute in which the penalty was taken. We also know the goalkeeper's and penalty taker's age and how long they had been playing in the league.

Converting a penalty towards the end of the match might be more difficult than towards the beginning of the match since the pressure to score a goal towards the end of the match is often higher. Older players tend to be more experienced which, in turn, may make them more self-confident and successful in penalty situations. Other relevant information may be the current score of the match and which team is playing at home; there may be a home advantage for penalty takers or goalkeepers and taking a penalty with a three goal lead may be easier than with one goal down.

Although we can collect a lot of different data on the Bundesliga penalties, it is difficult to make statements based on some of the information. For example, the variable 'club' does not have a natural order, which makes it hard to put a number on it and, in turn, analyze an influence or examine a trend based on those numbers. On the other hand, there are variables we could analyze but which are hard to obtain. Among these are the height of the shot. Is it easier for goalkeepers to save penalties for high shots compared to low shots? These data are not easily available, especially for older matches which were not recorded or broadcast on TV.

9.3 Factors Associated with the Outcome of the Penalty

9.3.1 Modeling Penalty Probabilities

In this section we will explain how we can determine effects that influence the result of a penalty from our data set. For this purpose, regression analysis is a suitable method. A simple linear regression model describes a response or dependent variable via one or more explanatory or independent variables with the help of a linear equation. An example for this is describing the height of a person dependent on their body weight or dependent on their parents' heights. For one explanatory variable we can display this dependency by a regression line. Figure 9.2 shows such a line for regressing goalkeepers' experience in the Bundesliga (measured in number of seasons played) on goalkeepers' age. In a regression analysis, we can also check how well the explanatory variables describe the response variable and whether a linear relation can be assumed at all.

We want to model the outcome of a penalty by means of different explanatory variables. The response variable is the outcome of the penalty, i.e. whether the penalty was converted or not. Because we want to use the ability of the goalkeeper as explanatory variable, it is sensible to only include penalties that have either been converted or saved by the goalkeeper. We therefore exclude penalties that have missed the goal, even though there may be good reasons to include them in other

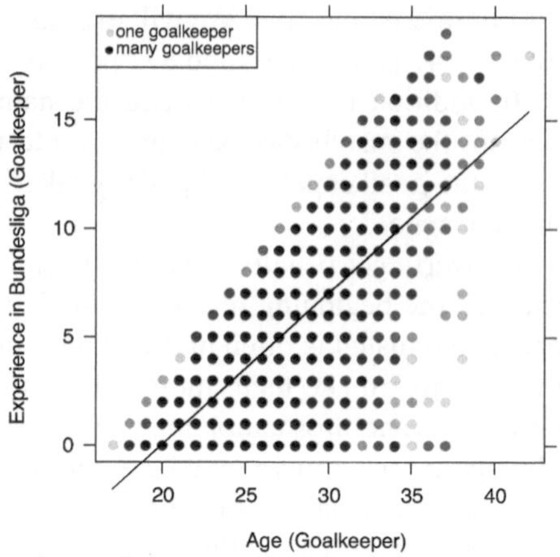

Fig. 9.2 Regression between goalkeepers' experience measured in number of seasons in the Bundesliga and the goalkeepers' age with corresponding regression line

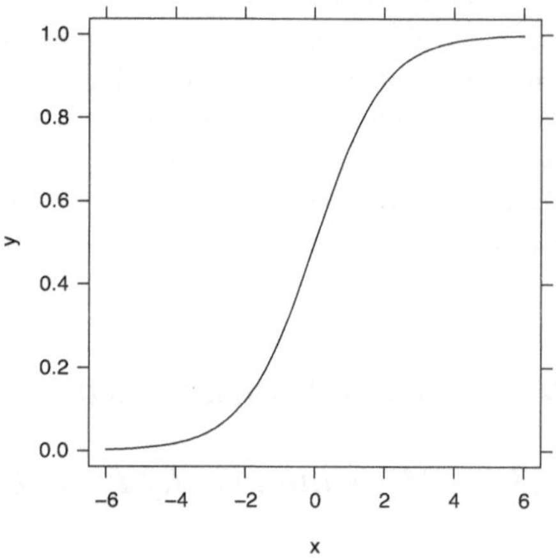

Fig. 9.3 Logistic function

contexts. Our response variable can only take two values: either the penalty was converted or saved by the goalkeeper. This can be coded into the model using 1 and 0, respectively. When analyzing a response variable with two possible values (a dichotomous response variable), a logistic regression model is a common choice. Instead of a straight regression line as in Fig. 9.2, a logistic curve is employed to model the response variable. This curve takes values between 0 and 1, an example is shown in Fig. 9.3.

We model the outcome of a penalty by means of a total of nine explanatory variables. This means, we will include nine explanatory variables in our Model 1 and analyze their influence. Apart from the goalkeeper's ability, we will also employ home court advantage for the goalkeeper, season, match day, minute of play, score at the time of the penalty, as well as goalkeeper's and penalty taker's age. In contrast to the goalkeepers' ability, we will not directly include the penalty takers' ability. Instead, we use the logarithm of the total number of penalties in their career as a proxy for their ability in Model 1.

9.3.2   Fixed and Random Effects

With the exception of the goalkeeper's ability, all variables are directly based on the data gathered. In statistics, the effects of such variables in a regression model are called fixed effects. The goalkeeper's ability cannot easily be quantified by raw data as such. For that reason, this variable exhibits uncertainty. Statisticians model this uncertainty via a random distribution that suitably describes the variable in question, in our case, the goalkeeper's penalty ability. The effect of such a variable in a regression model is called a random effect. As random distribution, often the normal distribution is chosen. The normal distribution is also known as Gaussian distribution or bell curve, because its probability density function has the shape of a symmetric bell. The center of the bell curve and the highest point of the probability density function is the arithmetic mean. Additionally, the standard deviation is important for normal distributions, because it describes the variance or dispersion around the mean. Values further away from the mean exhibit a lower probability. If we were to draw random numbers from a normal distribution, we would draw values close to the mean with high probability, while values more than two standard deviations away from the mean, e.g., would be drawn with low probability.

On the left-hand side of Fig. 9.4, we can see that the goalkeepers' abilities are not symmetrically distributed around the mean. In such a case, modeling the random effect via the symmetric normal distribution would not lead to a good model. Instead, statisticians employ a sum of multiple normal distributions, where the number of distributions is relatively low (e.g., 2 or 3). This number can either be chosen based on the data or by introducing an additional random component. The latter leads to a model called Dirichlet process mixture model. A regression model with both random and fixed effects is called a mixed regression model. Our Model

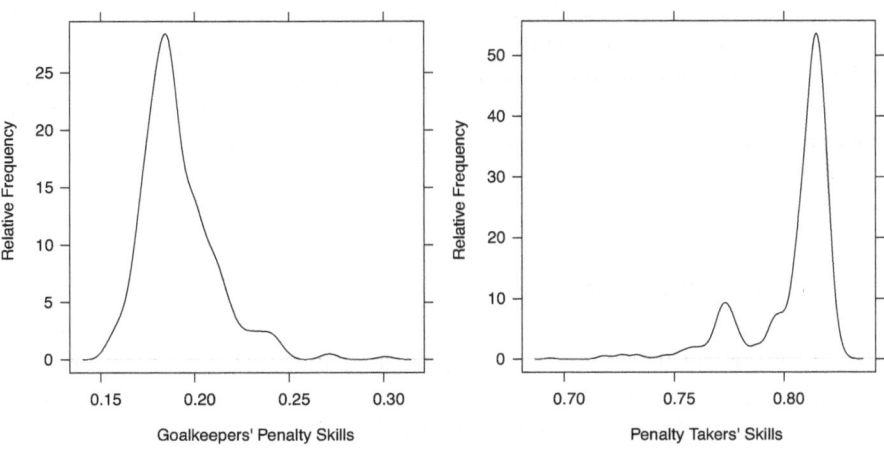

Fig. 9.4 Relative frequencies—based on Model 2 (cf. Sect. 9.4)—of the goalkeepers' skills (left-hand side) and penalty takers' skills (right-hand side)

1 is an example for a mixed logistic regression model with eight fixed effects and one random effect.

As we have seen in Fig. 9.1, there are only a few goalkeepers that we know a lot about, because they were involved in many penalties. On the other hand, there are many goalkeepers that we do not know a lot about. In these cases, we want to include additional information drawn from comparable goalkeepers. In statistics you can achieve this by including prior information in the form of so-called prior distributions for the variables. The analysis of models with prior distributions is called Bayesian analysis. Our Model 1 is an example for a mixed Bayesian logistic regression model with eight fixed effects and one random effect.

In order to analyze this Bayesian model, an algorithm is used that evaluates the model multiple times, each time taking a different draw from the random effect. To obtain good estimates especially for the random effects, the algorithm will be evaluated several thousand times. For the analysis of Model 1, we used 5000 iterations. In each iteration, the so-called regression coefficients are calculated, which describe the effects of the explanatory variables on the response variable, i.e. on the outcome of the penalty. A positive or negative value indicate a positive or negative influence, respectively. In our case, a positive influence of the goalkeeper's ability can be interpreted as a higher probability of saving the penalty. In contrast, a negative influence stands for a lower probability of saving the penalty, which means the probability of converting the penalty is higher. Evaluation and calculation of the model was done in the statistical software packages R (www.r-project.org) and OpenBUGS (www.openbugs.net).

To come to a statistically sound conclusion on the strength of the effects, we look at the so-called quantiles of the regression coefficients. In particular, we look at a lower and an upper quantile. We choose these quantiles in such a way that only few values are located outside of their range. For that reason, often the 2.5% and the 97.5% quantiles are chosen, which means that 2.5% of the lowest values are below the lower and 2.5% of the highest values are above the upper quantile. The interval of lower and upper quantile thus includes the majority of 95% of the values. If the values of both quantiles are positive or both are negative, i.e. the signs of the quantiles are the same, we can assume a meaningful influence of the effect on the outcome of the penalty. If the signs of the quantiles are different, no meaningful influence can be found.

9.3.3 Important Factors

In Sects. 9.3.1 and 9.3.2, we introduced Model 1 to answer questions regarding the importance of the fixed and random effects on the outcome of the penalties. Table 9.1 shows the results of this model. We can see that of all effects in Model 1 only penalty takers and season show the same signs for the upper and the lower quantiles. This means that for all other effects there is no meaningful (significant) influence on the outcome of the penalty. For penalty takers and

Table 9.1 Strength of the effects on the outcome of a penalty, according to Model 1. Meaningful influence factors are printed in bold

	Average	2.5% Quantile	97.5% Quantile
Penalty taker	−0.291	−0.345	−0.239
Home court advantage	−0.056	−0.226	0.109
Season	−0.007	−0.013	−0.001
Match day	0.001	−0.007	0.008
Minute of play	−0.001	−0.004	0.003
Score	0.038	−0.040	0.115
Goalkeeper's age	0.001	−0.017	0.020
Penalty taker's age	0.011	−0.011	0.034

season, both quantiles are negative, which indicates a negative influence of both variables on the outcome of the penalty. A penalty taker with a high number of total penalties thus exhibits a higher probability of converting the penalty, while, correspondingly, the goalkeeper shows a lower probability of saving the penalty. In order to see how much this probability changes on average if the penalty taker is more experienced, we apply the exponential function to the average effect −0.291, leading to $\exp(-0.291) = 0.75$. The probability for a goalkeeper to save a penalty is multiplied by 0.75 when the penalty taker has taken two instead of one penalties (measured on the logarithmic scale). Season also shows a negative effect, which means that it gets harder for the goalkeeper to save a penalty over the course of the seasons. However, this effect is small, because both quantiles——while negative— are close to 0.

9.4 Rankings of Goalkeepers and Penalty Takers

9.4.1 Penalty Takers' Influence as Random Effect

Model 1 yields a sensible ranking for the goalkeepers' penalty skills, but not for the penalty taker. Because we employ the logarithm of the total number of penalties as proxy variable for the penalty takers' skills, the taker with the highest number of total penalties would be considered the strongest in Model 1. This would rank the penalty takers according to their total number of penalties taken, but not according to their underlying ability.

In order to also model the penalty taker's effect in a way that leads to a meaningful ranking, we introduce a random effect instead of the fixed effect of the logarithmic number of total penalties. This random effect of the penalty taker's skill is modeled in analogy to the goalkeeper's penalty skill. To take into account that a penalty always constitutes a duel between goalkeeper and penalty taker, we combine the goalkeeper's and the penalty taker's effects into one model. Because there are fewer goalkeepers than penalty takers, we nest the penalty taker's effect within the goalkeeper's effect. The nesting ensures that only duels that have actually taken place are included into the model.

The strong penalty taker's effect that we can see in Table 9.1 justifies the high effort required for adding this further random effect. Since we have learned from Table 9.1 that besides the penalty taker's effect season is the only other meaningful variable, we only consider season as fixed effect. Model 2 is an example for a mixed Bayesian logistic regression model with one fixed effect and two random effects. As was the case with Model 1, we use 5000 iterations to calculate Model 2.

9.4.2 Leaderboards

The results of the penalty skills for goalkeepers and penalty takers based on Model 2 can be seen in Fig. 9.4; on the left-hand side, the goalkeepers' skills are shown, those of the penalty takers are illustrated on the right-hand side. We can see that most goalkeepers exhibit a penalty skill of just under 20% while the penalty skill of most penalty takers is over 80%. The tails are of special importance, because they contain the goalkeepers and penalty takers with the best and worst penalty skills. We can see that there are more unusually high outliers for the goalkeepers and more unusually low outliers for the penalty takers.

Tables 9.2 and 9.3 show excerpts of the leaderboards based on Model 2. These contain the best and worst of the 347 goalkeepers and 871 penalty takers, respectively, as well as some players of specific interest for soccer fans. In principle, players who have taken part in a few penalties only are sorted into the middle of the leaderboards by the regression model, because there is little information available

Table 9.2 Ranking of the goalkeepers' penalty skills based on Model 2. 'rel. f.' stands for relative frequency, 'sav.' for saved and 'tot.' for total number of penalties. 'Club' indicates the club the player played most matches for

Goalkeeper	Club	Seasons	Rank	Mean with quantiles	rel. f.	sav./tot.
Kargus, R.	Hamburg	1971–1987	1	0.241 [0.166, 0.332]	0.329	23/70
Enke, R.	Hannover	1998–2010	2	0.241 [0.156, 0.357]	0.407	11/27
Pfaff, J.	B. München	1982–1990	3	0.239 [0.141, 0.376]	0.545	6/11
Köpke, A.	Nürnberg	1986–1999	4	0.230 [0.154, 0.330]	0.333	14/42
Trapp, K.	Frankfurt	2008–2017	34	0.211 [0.128, 0.320]	0.333	5/15
Neuer, M.	S04/B.M.	2006–2017	40	0.210 [0.133, 0.311]	0.300	6/20
Illgner, B.	1. FC Köln	1985–1996	193	0.183 [0.112, 0.267]	0.172	5/29
ter Stegen, M.	M'gladbach	2010–2014	289	0.175 [0.093, 0.271]	0.083	1/12
Schmadtke, J.	Düsseldorf	1985–1997	344	0.158 [0.091, 0.228]	0.091	4/44
Junghans, W.	Schalke 04	1979–1991	345	0.156 [0.084, 0.234]	0.040	1/25
Müller, M.	Wuppertal	1972–1987	346	0.155 [0.080, 0.234]	0.040	1/25
Zieler, R.	Hannover	2010–2016	347	0.154 [0.074, 0.237]	0.000	0/19

Table 9.3 Ranking of the penalty takers' skills based on Model 2. 'rel. f.' stands for relative frequency, 'con.' for converted and 'tot.' for total number of penalties. 'Club' indicates the club the player played most matches for

Player	Club	Seasons	Rank	Mean with quantiles	rel. f.	con./tot.
Ritschel, M.	Offenbach	1970–1978	1	0.825 [0.770, 0.898]	1.000	17/17
Nolden, L.	Duisburg	1963–1967	2	0.823 [0.769, 0.890]	1.000	15/15
Heiß, A.	1860 München	1963–1970	3	0.829 [0.718, 0.996]	1.000	2/2
Ya Konan, D.	Hannover	2009–2015	4	0.826 [0.717, 0.945]	1.000	2/2
Götze, M.	BVB/B.M.	2009–2017	137	0.818 [0.648, 0.996]	1.000	1/1
Müller, T.	B. München	2008–2017	202	0.817 [0.758, 0.877]	0.929	13/14
Reus, M.	Dortmund	2009–2017	307	0.817 [0.751, 0.884]	0.900	9/10
Müller, G.	B. München	1965–1979	424	0.815 [0.772, 0.857]	0.810	51/63
Labbadia, B.	B. M./others	1987–2000	868	0.780 [0.636, 0.858]	0.200	1/5
Elmer, M.	Stuttgart	1973–1983	869	0.753 [0.509, 0.871]	0.000	0/2
Borowski, T.	Bremen	2000–2012	870	0.753 [0.521, 0.869]	0.000	0/2
Becker, E.	Karlsruhe	1980–1985	871	0.694 [0.223, 0.895]	0.000	0/1

about them and they are regressed to the mean by the model. The bottom and the top of the leaderboards mainly include players who have taken part in many penalties, because the abundance of data leads to more reliable evidence about them. This is particularly true for the goalkeepers. As illustrated in Table 9.2, Rudolf Kargus, who played for Hamburger SV for a long time, leads the ranking, while Ron-Robert Zieler, who did not save a single penalty until the end of the season 2016/2017, is last. Manuel Neuer, long-time goalkeeper of the German national team, is ranked at number 40 and thus quite far up. Other former or current goalkeepers of the German national team like Kevin Trapp, Bodo Illgner and Marc-André ter Stegen show mixed results with Trapp being the highest up in the leaderboard. In contrast to Neuer, the latter three played multiple seasons in France or Spain. Our analysis only considers matches in the German Bundesliga; their performance abroad thus has no influence on the rankings shown in Table 9.2.

As shown in Table 9.3, the penalty takers at the top of the ranking were all able to convert 100% of their penalties, whereas penalty takers with no converted penalties can be found towards the bottom of the list. In the history of the German Bundesliga, Gerd Müller is the penalty taker with the highest number of attempted penalties. However, due to his relative frequency of converted penalties, he finds himself in the middle of the ranking. The differences among the goalkeepers and among the penalty takers are rather small. This can be seen by comparing the intervals given by the 2.5% and 97.5% quantiles in Tables 9.2 and 9.3. These overlap for the best and worst goalkeeper ([0.166, 0.332] and [0.074, 0.237], respectively) as well as for the best and worst penalty taker ([0.770, 0.898] and [0.223, 0.895], respectively).

9.5   Conclusion and Outlook

In this chapter we have shown how the penalty skills of goalkeepers and penalty takers can better be estimated by statistical regression models, which has advantages compared to just reporting relative frequencies. We also illustrated how the influence of additional factors on the penalty outcome can be analysed. It has become obvious that a complex statistical model is needed to obtain reliable rankings. In addition to the influential factors considered here, the influence of further factors on the outcome of a penalty can be investigated given that corresponding data are available or can be collected.

Our investigations only take into account penalties from games of the German Bundesliga. The analysis can be transferred, of course, to games of other national soccer leagues as well as to international competitions for clubs or national teams. There is a special appeal of penalty shoot-outs in international competitions, where penalty shots have an even greater importance. With the video referral system that is used more and more in leagues and tournaments, the number of penalties seems to increase. An indication for this are the 22 penalties in the 2022 World Cup and the 29 penalties in the 2018 World Cup. Both exceeded the former record number of 18 penalties in the world cups of 1990, 1998, and 2002.

9.6   Further Reading

The contents of this chapter are based on the 2017 master thesis (in German) on modelling the penalty skills of goalkeepers and penalty takers by Peter Gnändinger, the first author of this chapter. This thesis, in turn, builds on the 2009 article on "Penalty Specialists Among Goalkeepers – A Nonparametric Bayesian Analysis of 44 Years of the German Bundesliga" by Björn Bornkamp, Arno Fritsch, Oliver Kuss, and Katja Ickstadt. The master thesis contains a detailed description of the underlying statistical method and a complete ranking for the goalkeepers as well as for the penalty takers based on the full statistical analysis. For further reading on the topic of regression analysis, which forms the basic modelling framework of this chapter, we recommend the 2013 book "Regression: Models, Methods and Applications" by Ludwig Fahrmeir, Thomas Kneib, and Stefan Lang.

Chapter 10
Music Data Analysis

Claus Weihs

Abstract This chapter illustrates how music audio data can be statistically analyzed for automatic transcription and determination of genre, with both processes based on signal-based features of audio recordings of musical pieces.

10.1 What Is Music?

For many people, music takes up large parts of their free time. The internet provides a number of different search engines for every taste, that work on the basis of tags, often genre classifications, that users have determined. However, how do genres differ from each other? What, for example, distinguishes metal rock from pop music, and how is classical music different from modern pop? Which sound characteristics are decisive for such a distinction? Do universally valid criteria exist? We will return to these questions at the end of the chapter. In addition, we will try to fulfil the dream of every musician, i.e. to automatically transform audio into notes. This will be done by means of statistical analyses. However, what does music have to do with mathematics and statistics?

Scholars have long puzzled over the intense emotional effects of music. Way ahead of his time, the philosopher Pythagoras was the first to describe an intriguing connection between mathematics and music. Around 500 BC, he examined the structure of music with the help of a monochord, a kind of guitar with only a single string (cf. Fig. 10.1). He realized that the basic musical intervals can be described by simple numerical ratios. He showed this by repeatedly shortening the string of the monochord. When shortened to 3/4 of its original length, plucking the string produces a tone that is a fourth higher than the original, shortening to 2/3 leads to a fifth and shortening to 1/2 to an octave higher than the unshortened string

C. Weihs (✉)
TU Dortmund, Department of Statistics, Dortmund, Germany
e-mail: claus.weihs@tu-dortmund.de

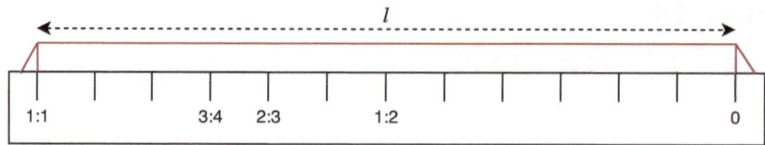

Fig. 10.1 Monochord with string of length *l* and shortening ratios

(cf. Fig. 10.1). Pythagoras also developed the world's first musical scale, that has undergone slight modifications but is still valid and in use in Western music today.

It was not until the seventeenth century, long after the invention of music notation, polyphony, and harmony, that the French monk and mathematician Mersenne succeeded in giving a physical explanation to the findings of Pythagoras. Mersenne made strings sound that were up to 40 m long. He counted their vibrations and was able to confirm Pythagoras' finding that an octave always vibrates twice as fast as the respective fundamental tone. What distinguishes music from noise is the regularity of the vibrations, i.e. of the back and forth movement of air molecules. Tones usually consist of vibrations of several overlapping frequencies, i.e. the fundamental tone and the corresponding overtones. The fundamental tone determines the pitch we perceive, and the overtones sound with multiples of its frequency and represent the timbre.

Considering the complexity of a single tone and the number of instruments in an orchestra, it is an amazing achievement of the human sense of hearing that we are able to process the overall sound pattern in a concert hall and, at the same time, to filter out every single sound. Yet, the ear is a comparatively poorly equipped organ. With about 3500 sensory hair cells only (for comparison: the eye has 120 million photoreceptor cells), it transforms sound waves into electrical impulses (cf. Fig. 10.2). The tiny fluctuations in air pressure produced by the sound wave are received by the eardrum (tympanic membrane), amplified via the auditory ossicles (malleus, incus, stapes), and transmitted to a membrane at the entrance (oval window) of the fluid-filled inner ear. The snail-shaped sensory organ (cochlea) then splits the incoming sound into its individual frequencies. Low-pitched sounds make their way deep into the cochlea where they are converted into nerve impulses. High-pitched sounds, on the other hand, are converted to nerve impulses close to the entrance to the inner ear. By this mechanism, the ear is able to distinguish even sounds that are only 1/10 of a semitone step apart.

In ancient Greece, muses were the goddesses of inspiration in literature, science, and art and were considered the source of all knowledge. The Latin word musica and our word music are derived from the Greek word mousike, meaning 'the art of the muses'. Unfortunately, there is no universally accepted definition of the notion 'music'. Therefore, it is often approached through its components. The basic musical signals are tones. Tones, in turn, have four dimensions, namely pitch, loudness, duration, and timbre, that are introduced in the following. Tones are distinguished by their names (c, d, e, f, ...), that correspond to their pitch. Two tones

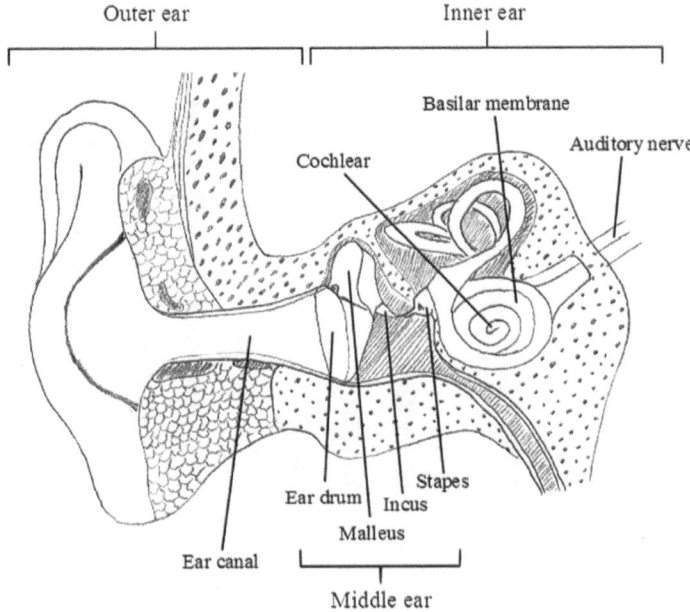

Fig. 10.2 The human ear (Copyright 2017 from Music Data Analysis: Foundations and Applications, Figure 6.1, p. 166, by Claus Weihs, Dietmar Jannach, Igor Vatolkin, Günter Rudolph. Reproduced by permission of Taylor and Francis Group, LLC, a division of Informa plc.)

of the same pitch but different loudness or timbres are notated as the same tone but played with different intensity and possibly on different instruments. Timbres are created by the intensity of the overtones of the fundamental, whose frequencies are multiples of the frequency of the fundamental. The frequency of the fundamental is perceived as pitch, the intensities of the overtones as timbre. Different instruments can generally be distinguished by the strength of their overtones. Figure 10.3 shows an example of the distribution of overtone intensities for piano and electric guitar. OT0 corresponds to the fundamental, OT1 to the first overtone, etc.

10.2   Music Data

We are interested in the sound properties pitch, volume, duration, and timbre. Unfortunately, these properties are generally not readily available in audio recordings and need to be extracted first. Generally, this information is elicited for short time windows, e.g. of 4 seconds, so that changes in the properties can be quickly detected by comparison with previous time windows. To this end, we first determine the spectrum, i.e. the strength of the different frequencies in the time windows, from that so-called audio characteristics are derived. These, in turn, represent the sound properties we are interested in.

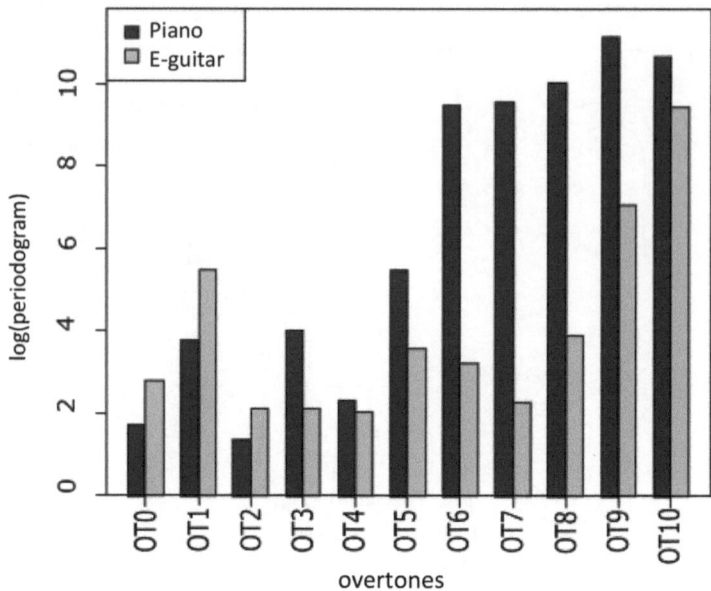

Fig. 10.3 Strength of the overtones (OT0 = fundamental, OT1, . . . , OT10) for piano and guitar on logarithmic scale

In the following, we provide examples of audio characteristics from that onset, harmony, and timbre can be derived. Hundreds of such audio characteristics are proposed in the literature, all of which are more or less suitable for determining musical tone properties. We will concentrate on some of the most important ones. One rhythmic feature is 'Spectral Flux', that is defined as the magnitude of spectral change between two adjacent time windows, and is an adequate characteristic to successfully identify tone onsets (cf. Sect. 10.3). An important harmonic feature is the 'Chromagram', that represents the spectral energy of all 12 semitones of the Western musical scale for each time window; octaves are ignored. The Chromagram can, for example, be used to identify keys. The most widely used features for identifying instruments (by means of timbre) and genres are the so-called 'Mel Frequency Cepstral Coefficients' (MFCCs) (cf. Sect. 10.3). The main function of these coefficients is to convert the acoustic frequencies of musical tones into perceived pitches, i.e. the so-called mel scale (mel from the word 'melody'). After a number of further transformations not introduced here, MFCCs describe both the general shape of the spectrum and the spectral fine structure.

In addition to these signal-based features further information exists about music tracks. For example, we can obtain information from music websites. Examples are so-called tags (labels) that refer to the genre of a piece of music or the feelings and emotions listeners might have when consuming the piece of music. A plethora of playlists that are associated with such genres or emotions have been created by individual users. Music services such as Last.fm offer users the possibility to tag pieces of music for easier access to similar pieces. Various music databases exist online that contain metadata for millions of songs. Since these are not of immediate relevance for our current analyses, we will focus on signal-based features in the following.

10.3   The Studies

In the following, we present the results of exemplary studies in music data analysis for pitch determination, instrument recognition, onset time recognition, automatic transcription, and genre determination. Before we do so, we introduce the statistical method employed, i.e. classification modeling.

10.3.1   *Classification*

In classification, we analyze the relationship between a so-called class variable (e.g. musical instrument) and those variables that have a potential influence on such a class variable (e.g. signal-based features). The aim is to predict the true value of the class variable (e.g. piano or guitar) from the values of the influential variables. In order to determine a prediction model, the true class together with the respective values of the influential features must be observed on n different subjects/objects ('training sample'). From these observations, a 'classification rule' is derived that can be used to predict an unknown class from known values of the influential features. The simplest classification problem involves only two classes, e.g. the two musical instruments piano and guitar, the genres classical and non-classical, or the question whether an onset exists in a time window (yes/no). In pitch determination, however, the classification problem has more than two classes (e.g. the 12 semitones). The validity of classification rules is typically assessed by error rates, i.e. the number of errors in relation to the number of observations.

10.3.2   *Pitch Identification*

Nowadays, pitch is identified by the frequency of a sine wave. This idea dates back to the late eighteenth century, e.g. to Daniel Bernoulli (1700–1782). In 1822, Fourier presented his famous theorem that periodic functions can be represented by using sums of sine and cosine functions. Subsequently, Ohm (1843) applied this theorem to acoustics. Since then, the frequencies of sine waves have been used as a measurable representation of pitch (cf. Fig. 10.4).

The frequency of a sound can be identified, for example, by means of a so-called auditory model. This way, a simple mathematical/statistical model approximates the

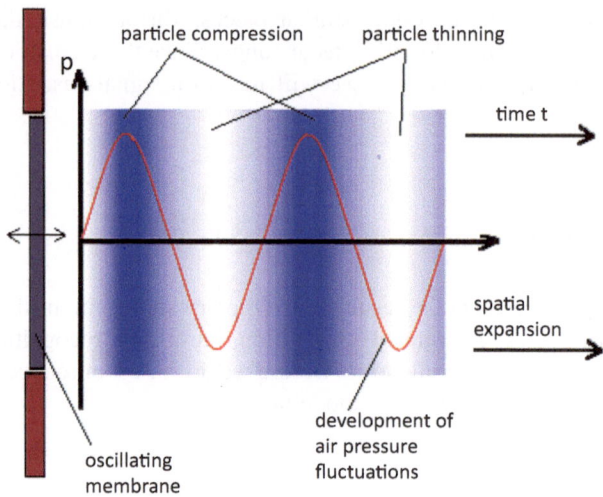

Fig. 10.4 Sine model for sound waves

properties of the human ear. Based on animal observations and other psychoacoustic phenomena, it appears reasonable to identify the frequency of sounds on the basis of 30 channels (instead of about 3500 (inner) hair cells in the human ear), that result from partitioning the hearing range. Each channel has a specific best frequency (BF) between 100 and 3000 Hz. BFs are defined by the frequencies that are most stimulated in the respective channel. In our study, we use individual classification models for each channel to decide whether that channel represents the fundamental frequency or the overtones of the sound.

For example, a simple rule could read as follows:

If the strongest frequency in a channel has a very high spectral energy and the spectral energy of the next strongest frequency is low, the strongest frequency can either be identified as the fundamental frequency or one of the overtone frequencies of the sound.

The identified frequencies are then put into relation with each other in order to determine the frequency of the fundamental, i.e. the pitch. Our results have shown that, by combining several simple rules for each channel, it is possible to determine the correct frequency with very small error rates. In Table 10.1, the channels with the smaller numbers represent the lower frequencies. We can therefore conclude that pitch estimation is more accurate for lower frequencies since they come with smaller error rates.

Table 10.1 Error rates in the different channels of the auditory model

Channel	1	2	3	4	5	6	7	8	9	10
Error rate in %	0.2	0.4	0.5	0.5	0.3	0.3	0.3	0.4	0.6	0.7
Channel	11	12	13	14	15	16	17	18	19	20
Error rate in %	0.6	0.5	0.7	0.7	0.6	0.9	1.0	1.2	1.4	1.3
Channel	21	22	23	24	25	26	27	28	29	30
Error rate in %	1.4	1.3	1.4	1.4	1.3	1.4	1.3	1.1	1.0	1.1

10.3.3   Instrument Recognition

Two musical notes may sound different even though they have the same pitch, volume, and duration. One reason for this might be that they were played by different instruments, e.g. a piano or a trumpet. Differences between two tones that are not due to pitch, volume, and duration are referred to as timbre. Note, however, that notes played by the same instrument but with different pitches or volumes can also have different timbres. No fixed or constant timbre exists for the individual instruments.

The timbre of a sound is essentially determined by its spectrum, i.e. by the intensity of the various overtones and additional noise components, as well as by the temporal development of a sound, in particular around its onset. To quantify timbre, one therefore needs to determine the temporal development of the spectrum.

In one of our studies, we aimed to identify particular instruments on the basis of their timbre. For example, we looked into which sound characteristics were crucial to distinguish piano and guitar tones from each other. To this end, we determined a classification rule, that assigns an instrument to certain sound characteristics. As an example for these characteristics, we used the first 13 MFCCs (cf. Sect. 10.2), that we have measured for different parts of each tone. Based on these characteristics, a rule was determined from 4309 guitar and 1345 piano tones. The rule had an error rate of approximately 5.5% only. The results have shown that the MFCCs of the beginning and the end of the tones were more important for instrument recognition than of the middle of the tones. This means that initial tone production, i.e. plucking in the case of the guitar and hammer stroke in the case of the piano, and the decay characteristics of the tones were more important for distinguishing the two instruments than the full manifestation of the sound in the middle of the tones.

10.3.4   Onset Detection

In another study, we investigated onset detection. To this end, we used an 'onset detection function' (ODF) and determined its local maxima for the progression of a sound. A local maximum is a time point in the progression that has a higher ODF

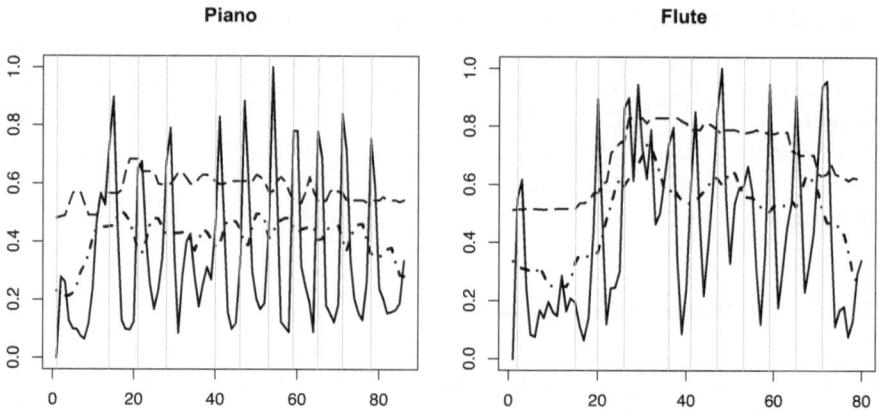

Fig. 10.5 Different thresholds (indicated by different line types) for Spectral flux as the ODF applied to piano (left) and flute (right) over time. The true onsets are indicated by vertical lines

value than all other time points in its vicinity. Such local maxima constitute onsets if their ODF values are higher than a pre-defined threshold. We varied these thresholds over time to react to variation in loudness that, e.g., might be caused by the dynamics of musical pieces (e.g., piano, forte). In order to not overlook an onset, the signal is divided into small (overlapping) time windows on that the ODF is calculated and local maxima are identified. If a local maximum is larger than the threshold in any of these windows, the beginning of a new tone is assumed (onset).

Thresholds can be set in different ways. We used thresholds which guarantee that local maxima which are identified as onsets are greater than a linear function of the mean or median of the ODF in the neighborhood of the local maximum. In Fig. 10.5 you can see two versions of the threshold represented by dotted and dashed lines, respectively. We used Spectral Flux (cf. Sect. 10.2) as the ODF and applied it to sounds of piano and flute. Figure 10.5 shows how important it is to choose the right threshold. With the threshold represented as a dashed line, some onsets of the flute are not recognized. With the threshold represented as a dotted line, on the other hand, some onsets are wrongly recognized. Furthermore, Fig. 10.5 shows that onset detection with Spectral Flux as the ODF works very well for instruments with a well-defined tone onset (like piano). For other instruments, the beginning of the tone may even not be well-defined (like for flute).

10.3.5 Automatic Transcription

For the automatic transcription of audio tones to musical notes directly from the progression of sound, various tasks have to be solved. In particular, the onsets of the different tones must be determined and thus the tone durations (cf. Sect. 10.3.4) as

Fig. 10.6 Excerpt from 'Tochter Zion' (Daughter of Zion): (**a**) original notes, (**b**) tuneR transcription, and (**c**) Melodyne transcription of a vocal recording

well as the pitches of the tones with the identified onsets (cf. Sect. 10.3.2). Moreover, the identified onsets must be fitted into bars, i.e. they need to be 'quantized'. Before that, it must be clarified which musical meter (time signature such as 3/4 or 4/4 time) applies to the piece of music. Finally, in polyphonic pieces of music, simultaneously played tones must be identified and these tones must be assigned to instruments. For this purpose, instrument recognition can be helpful, for example, to identify whether an instrument is playing or not (cf. Sect. 10.3.3).

The transcription of monophonic pieces of music into musical notes is already quite well understood. We studied automatic transcriptions of the song 'Tochter Zion' (Daughter of Zion). In Fig. 10.6, one can see two transcriptions of an excerpt of this song recorded from a performance by a professional female singer. The transcription in (b) was generated by our software tuneR and the transcription in (c) by the commercial software Melodyne. When compared with the original in (a), both scores obviously are not optimal. However, it may well be that the singer, who performed the song, did not fully adhere to the original composition.

10.3.6   Genres

In another study, we considered a two-class example of genre recognition, namely the distinction between classical and non-classical music that included pop, rock, jazz, and other genres. As influential variables, we used the mean and the standard deviation of the first 13 MFCC features on 4-second time windows with 2-seconds overlap. We trained a classification rule on 10 classical and 10 non-classical music tracks. This resulted in 2361 observations across all the 20 pieces of music (training sample). The derived classification rules were tested on 15 classical and 105 non-classical music pieces (test sample). There was no overlap between music pieces and artists in the training and test samples.

The genre was predicted by the classification rules in small time windows of 4 seconds length. The typical error rate of the rules was 17% of the 15,387 small time windows of the test sample. However, for an application we were interested in how good longer sections of the pieces of music are misclassified. In fact, we found only 33 misclassified contiguous sections of 32 seconds length. In order to assess whether such a section is misclassified, we restricted ourselves to classification rules that determine probabilities for the classes in each time window. If the probability of the true class is less than 0.5, the window is misclassified. To identify contiguous misclassified parts of 32 seconds length, we used the mean of the probabilities of the true class over the small time windows building the longer part. It is striking that many of the misclassified parts with the smallest mean probabilities of the true class come from 8 pieces of European Jazz that was not represented in the training sample. Looking at the 5 smallest probabilities of the true class ($< 3.5\%$), only 2 of them were not from this group, namely 'Fake Empire' by the Indie Rock band The National and 'Trilogy' by Emerson, Lake, and Palmer. While 'Trilogy' had to be assumed to be near to classical music beforehand, in the case of Fake Empire the beginning of the piece was misclassified, where only piano and vocals are active. Overall, our classification rules led to interpretable results, that, however, clarified how important a well-considered training sample is and that some parts of pop pieces are definitely close to classical music.

10.4 Further Reading

Section 10.1 is very largely based on parts of the journal article "Musik, die Mathematik der Gefühle" (Music, the Mathematics of Emotions) by Philip Bethge, DER SPIEGEL No. 31/28.07.2003. The results in Sect. 10.3.2 can be found in "Statistics for hearing aids: Auralization" by Claus Weihs, Klaus Friedrichs, and Bernd Bischl in the book "Data Analysis Methods and its Applications" (J. Pociecha and R. Decker, Eds., Wydawnictwo C.H. Beck, 2012, 183–196). The results in Sects. 10.3.3–10.3.6 are based on Chaps. 12, 13, 16, and 17 of the book "Music Data Analysis: Foundations and Applications," Chapman & Hall, 2016, edited by the author of this chapter and by Dietmar Jannach, Igor Vatolkin, and Günter Rudolph from the Faculty of Computer Science at TU Dortmund University. There are many more examples of music data analysis in this book, e.g. on tempo recognition, characterization of emotions in music and automatic composition.

For a documentation of the R-package 'tuneR' mentioned in Sect. 10.3.5 see https://cran.r-project.org/web/packages/tuneR/tuneR.pdf. A current version of the program Melodyne also mentioned in Sect. 10.3.5 can be found in https://www.celemony.com/en/start.

Chapter 11
Statistics and Horse Race Betting: Favorites vs. Longshots

Martin Kukuk

Abstract A strong motivation for engaging in statistics are bets of all kinds. They involve money, chances, and speculating on potentially recurrent events. Especially horse race betting is an interesting research object since individual decisions in a real-world situation can be observed and, hence, analyzed in relation to the actual outcomes of the races. Therefore, betting is a kind of field experiment with many participants, in contrast to lab experiments where only a few participants make choices in controlled designs.

11.1 Horse Race Betting

The empirical analysis of horse races and especially the bets placed on them have a long tradition. The most important finding in almost all studies is that with bets on horses with low winning chances (longshots) more money is lost than on those with high winning chances (favorites). In other words, when measured against winning probabilities, bets on longshots are relatively too frequent when compared to bets on favorites. This phenomenon, called favorite-longshot-bias (FLB), has systematically been found in studies on different countries, horse races, types of bets, and periods of time.

For German harness horse races, the differences in average profits or losses for either favorite or longshot race participants are also well-documented. In Anglo-Saxon countries, bookmaker bets are commonly offered, which pay fixed quotas to bettors in case a bet is successful. In Germany and many other countries, pari-

The author would like to thank the editors Buschfeld and Weihs for their kind support to rewrite this chapter to make it accessible to a wider audience.

M. Kukuk (✉)
University of Würzburg, Department of Economics, Würzburg, Germany
e-mail: martin.kukuk@uni-wuerzburg.de

mutuel bets are widespread. For this type of bet, the sum of all bets are paid out to the successful bettors according to how much they bet on the winning horse (cf. Sect. 11.2)—after a fee has been deducted for the organizer (track take). It is common practice for winning bets that the payout for a one euro bet is constantly updated and announced right up to the start of the race. Due to the track take, the mean profit of a bet is negative. This is also true for bookmaker bets since bookmakers, too, want their effort and entrepreneurial risk to pay off. Nevertheless, both forms of betting enjoy great popularity.

11.2   Betting Payouts

In the following, pari-mutuel betting is considered more closely. This type of bet is particularly interesting for statisticians, since the multitude of bets placed shortly before the start of the race creates a complex, rather unpredictable interaction and outcome of numerous individual choices, which, at all times, are influenced by the constant updates on bets wagered so far.

Let us consider the following example: At a betting deadline, 100 bettors wagered one euro each. Of these 100 euros, 40 euros were bet on the win of a specific horse. If this horse actually wins (and one does not consider any track take), the 100 euros are paid out to the 40 successful bettors. Each euro successfully invested yields a payout (gross quota) of 100/40; the unsuccessful bets receive nothing. The inverse of this quota constitutes the bettors' subjective winning probability $p = 40/100$ for a specific horse, here the winning horse.

In our example, the bettor gets a payout of 100/40 with 40% probability, which yields a mean payout of one euro for this bet. Subtracting the one euro invested in this bet, a mean profit or mean return of zero is obtained. This constitutes a fair game if we neglect the deduction from the betting pool by the organizer before payout. Considering the track take results in a decrease in mean payout and consequently in a negative profit and thus in a loss.

For this type of bet, the bets on different horses yield the same mean payout per invested euro. However, if a longshot really wins, the payout will be much higher than the payout for a favorite. Since in both cases, the payout will be similarly negative if the horse does not win, the variance of payouts (i.e. the difference) between winning and losing is higher for longshots than for favorites. Therefore, the variance in the payouts increases with decreasing winning probabilities. The variance is usually considered to be a measure of how risky a bet is. This raises the question why a bettor, expecting the same mean payout for all horses, would take a higher risk by wagering on a longshot.

11.3 Empirical Explanations for the Favorite-Longshot Bias

The negative mean return is accepted by the bettors since, for some people, gambling is a free-time activity, similar to spending money on an opera ticket. Given the decision to spend a certain amount of money on horse race betting, the next question is on which horse to bet.

A successful bet on a longshot yields a higher payout quota compared to the lower payouts on favorites, which gives the bettor the opportunity to brag about this within his or her circle of friends and acquaintances. This is a rather rare outcome, which, however, generates a high admiration for the bettor. This provides a certain attraction for at least some race track gamblers and partly explains the favorite-longshot-bias.

But even for more rational bettors, betting on longshots constitutes an attractive option. Recall that if winning probabilities equal the shares of bets on a specific horse and the favorite horse has a 40% probability of winning, 100/40 euros will be paid out for each euro. The biggest longshot, however, may have a winning probability of 5% only, but would yield a much higher payout, i.e. 100/5 euros.

In the above scenario, which involves 100 bettors, if one of the gamblers who was initially in favor of the favorite horse changes his bet to the biggest longshot and thus the higher risk, the quota will change to 100/39 for the favorite and 100/6 for the longshot. Nevertheless, the general winning probabilities remain unchanged, i.e. 40% for the favorite and 5% for the longshot. If the percentage track take is 20%, the mean payout rate can be calculated to be -0.18 for the favorite and -0.33 for the longshot. The corresponding variances of the payouts between winning and losing would be 1.0 for the favorite and 8.4 for the longshot. Therefore, when the mean payout increases from -0.33 to -0.18, the variance decreases from 8.4 to 1.0. If depicted in a diagram in which the variance is shown on the y-axis and the mean on the x-axis (cf. Fig. 11.1), this leads to a decreasing curve.

Figure 11.1 illustrates some empirically observed relationships between mean relative returns and variance in payout from several races. The downward slope for wins (blue line) illustrates the result of the risk-loving behavior in the study. The left part of the line depicts the higher mean losses and higher variance for the longshots, whereas the right part of the line illustrates the corresponding lower values for the favorites.

In horse race betting, aside from win bets, bettors can also wager money on exactas and trifectas, i.e. bets in which bettors have to indicate the first two and the first three winners in their exact order, respectively. In yet another type of bet, called show bets, the backed horse must come in third, at least. As the negatively sloped curves in Fig. 11.1 suggest, the favorite-longshot-bias is also present in these types of bets. For exactas and trifectas, the curves are more erratic, since these types of bets are less frequent than the other types. In our example, this is particularly true since we consider, e.g., a very specific exacta bet in which the third favorite is set as the winner and the first favorite as the runner up. Such combinations are, of course, strongly affected by single outliers.

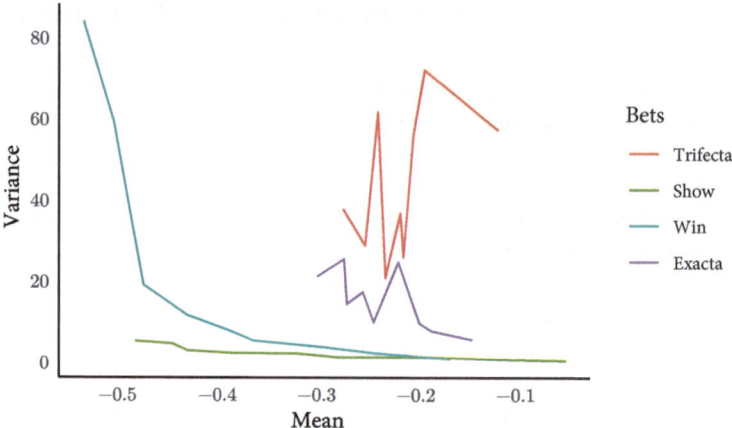

Fig. 11.1 Mean relative returns (in euros per invested euro) and variance (in squared euros per invested euro) of payouts for different types of bets for the same races

When comparing the curves in Fig. 11.1, we see that they are quite different from each other. If we have a set risk (variance) of 20 squared euros, mean losses are much lower for exacta and trifecta bets than for win bets. When taking the alternative perspective and determine an expected return of −20%, we see that the variance and thus the risk is higher for trifecta and exacta bets than for wins and shows.

11.4 Favorite-Longshot-Bias Caused by Subjective Estimates

A theoretical explanation for the favorite-longshot-bias is based on individually varying estimates of the winning probabilities, assuming that half the participants overestimate while the other half underestimate the unknown (objective winning) probabilities p. The median of the distribution of such estimates is the value for which half of the estimates are lower while the other half is higher. The median thus corresponds to the unknown probability p.

In our scenario of individually varying estimates, let us presume that bettors bet on the horse with the highest mean payout, since we assume that bettors neither seek nor try to avoid risk. The mean payout is determined by the individually estimated winning probability and the quota. For win bets, the pari-mutuel quota changes with every single bet and the share of bets on a particular horse is displayed at any given point in time. To estimate the probable payout for each horse, the bettor multiplies the current quota by its individual winning probability. If we assume that the bettor bets on the horse with the highest payout and, right before the betting deadline, the quota of the favorite is, for example, 100/70, i.e. the share of bets is b = 70/100, all bettors with an individual winning probability higher than or equal to 70% will bet on this horse. As a result, the share of bets b on this particular horse goes up and

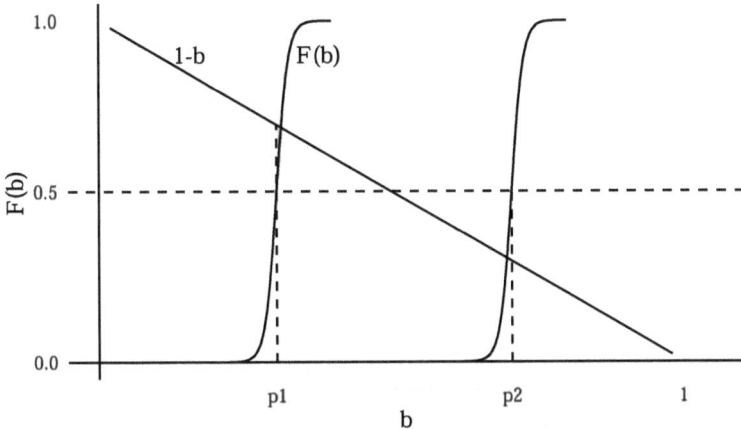

Fig. 11.2 Equilibrium share of bets: Distribution functions F(b) of the individual winning probability estimates against the share of bets b. At equilibrium, the share $1 - F(b)$ of individual estimates larger than b must be equal to the share of bets b, implying F(b)=1 − b. Horse 1 has a lower winning probability (p1) than horse 2 (p2)

ultimately equals the share of bettors with an individual winning probability higher than b (equilibrium share of bets).

For a statistical analysis, we consider the distribution of the estimates of the winning probabilities of each individual horse. The value of the distribution function F(x) of a distribution of such random values like our estimates is defined as the share of values of the distribution that are lower than or equal to x. To determine the share of values larger than x, we subtract F(x) from 100%, i.e. this share of values is $1 - F(x)$. In our case, the share of individual estimates of the winning probability larger than the share of bets b is $1 - F(b)$. Figure 11.2 displays the relationship between the share of bets b and the distributions of the individual estimates for two horses. The intersections of the falling line $1 - b$ with the distribution functions F(b) of the distributions of the individual probability estimates are equal to the equilibrium shares of bets, since F(b)=1 − b is equivalent to $1 - F(b)$=b. In Fig. 11.2, it can be seen that the horse with the lower winning probability (p1) has an equilibrium share which is slightly higher than p1. For the favorite with a winning probability p2, the equilibrium share is lower than p2. This shows that, in comparison to the objective winning probabilities, too many bets are placed on the longshot and too few on the favorite. This constitutes an economically motivated, theoretical approach towards the favorite-longshot-bias.

For the more realistic case of races with more than two horses, we can employ the same line of thinking. An individual bet on a particular horse is placed so that the estimated return is higher than for the other horses, i.e. the bettor tries to identify the horse with the highest winning probability. To come to a reliable conclusion, this needs to be done for all horses, which would result in a set of multiple equations. This can no longer be solved by means of economic theory. However, using Monte-

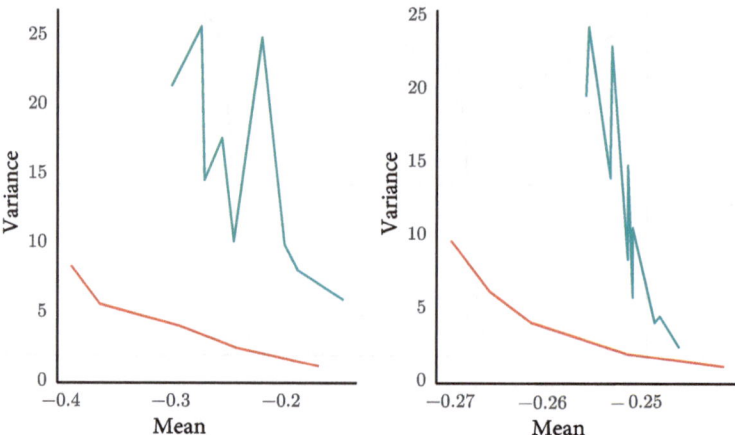

Fig. 11.3 Means and variances of returns of win and exacta bets of the same race. Real (left) and simulated (right) races

Carlo experiments, equilibrium quotas can be determined and hence mean-variance profiles. These profiles all show the typical decreasing curve, depending on the number of racing horses, the objective winning probabilities, and the amount of estimation uncertainty.

In such experiments, the other types of bets (i.e. exacta, trifecta, and show) can be considered as well if the whole race result is simulated. Let us assume that a risk neutral rational gambler does not only consider the mean returns from the win bets but additionally those of the exactas. The bettor then chooses the bet with the highest subjective mean return. As a result, different mean-variance profiles for win and exacta bets are obtained, which are displayed in the right part of Fig. 11.3. In the left part, the comparable empirical observations for win and exacta bets are displayed (cf. Fig. 11.1).

11.5   Conclusion

As the results in Sect. 11.4 have shown, economical approaches only help to explain the favorite-longshot-bias for win bets. If other types of bets are included in the analysis of betting behavior, the empirical results point to the fact that modern economic approaches are not sufficient for explaining this phenomenon. They typically rely on the assumption that rationally acting individuals know the objective probabilities. Also, experimental studies presuppose that probabilities are known in gaming situations. This does not apply to trifecta, exacta, and show bets, since on many national and international horse race tracks, the pari-mutuel quotas are only displayed and updated for win bets, which allows for a model for explaining the favorite-longshot-bias. For the other types of bets, an approach is needed which is able to factor in subjective estimates of probabilities on the one hand and the bettors' willingness to take risks on the other.

11.6 Further Reading

A selection of studies documenting the favorite-longshot-bias as well as other contributions explaining this phenomenon are presented in M. Kukuk and S. Winter (2008): "An Alternative Explanation of the Favorite-Longshot-Bias", The Journal of Gambling Business and Economics 2(2), 79–96.

Chapter 12
The Statistics of the German 6/49 Lotto

Walter Krämer

Abstract There is considerable evidence that positive returns from the lotto are possible in the long run. To many of us this seems paradoxical, since about 50% of revenues are retained by the state-owned lotto monopolies. But the other half is open to competition among players, where the smart ones may exploit the rest. Here we show how this works.

12.1 Lotto as an Investment

About half of all Germans play the lottery, at least occasionally. And—surprise, surprise—according to *Stiftung Warentest*, an organization established in 1964 by the German federal parliament with the aim of helping consumers by providing impartial and objective information, 93% of participants in gambling games play to win something. For most players, however, this ends in disappointment: Pure games of chance such as roulette, *Spiel 77*, which is only based on the lottery ticket number, or the well-known class lottery create negative individual returns in the long run. Otherwise, these games would be unprofitable for the organizers and cease to exist.

Things are different for strategic games of chance, where, in addition to chance, the behavior of the other players is also important for individual returns. A prime example is the game of lotto, which is widespread in different variants all over the world. The word lotto is taken from the Italian and French lot = share or fate. Contrary to what many believe, it is possible to win the lotto in the long run—i.e. not by pure luck alone, but on average if playing till the end of time.

The usual argument runs like this: Half of all stakes are collected by the state, leaving only 50 cents out of every euro to the gamblers—an average loss of 50 cents for every bet of one euro. And because the actual loss in the long run coincides with

W. Krämer (✉)
TU Dortmund, Department of Statistics, Dortmund, Germany
e-mail: walterk@statistik.tu-dortmund.de

© The Author(s), under exclusive license to Springer-Verlag GmbH, DE, part of Springer Nature 2024
C. Weihs et al. (eds.), *Statistics Today*, Society, Environment and Statistics, https://doi.org/10.1007/978-3-662-68907-3_12

101

the theoretical loss according to the law of large numbers, one can only lose in the long run.

This argument is false; it is only valid on average over all number combinations selected by all gamblers. Some combinations lose even more in the long run, others less. However, certain number combinations may actually generate profit in the long run. The only problem is to find them.

The usual algorithms for mean return only works for lotteries with fixed prizes for a single bet: Correctly choosing six numbers out of the winning combination six returns—say—three million euros, five correct numbers return 200,000 euros, and so on. The expected return is then the sum of products of the respective probabilities and returns. Both are fixed, and the sum of the products is always lower than the bet, so on average there results a loss. This is how the forerunners of the modern lottery were constructed, such as the famous Genoa 5/90 lottery, which still exists in Italy today. Here, a single correct number returned 14 times the bet, two correct numbers returned 240 times the bet, three correct numbers 4800 times the bet and four correct numbers returned 60,000 times the bet. No bets were accepted for five correct numbers.

In the modern 6/49 German *Zahlenlotto*, on the other hand, and in most similar formats elsewhere, only the probabilities of winning are fixed, but not the winnings (= prizes) themselves. These depend on the other bettors via the number of co-prize-winners. This lottery has been in existence since 1955, when the states of Bavaria, Hamburg, North Rhine-Westphalia, and Schleswig-Holstein founded the German *Lottoblock*. The first draw was held on October 9, 1955, a Sunday, in front of a live audience at the Mau Hotel in Hamburg; two orphan girls (orphans used to be very popular for such purposes) took turns to draw six balls out of an urn; the result was the sequence 13-41-3-23-12-16, making 13 the first lotto number ever drawn in Germany (and the rarest to this day).

The current division into nine prize classes and the distribution of the payout sum (which corresponds to exactly half of the bets) to these classes dates from 2013. In addition to six 'normal' balls, a seventh 'bonus ball' is now also drawn, and a fixed price of 5 euro is given to betters who have picked 2 winning numbers plus the bonus number. Balls are drawn live on TV every Saturday evening. A single bet costs 1.20 euro, plus an additional fixed 0.60 euro service charge. Table 12.1 shows the prize classes together with their probabilities.

No bettor can change the chances of winning. Finding these is not too difficult. Calculating the probability of picking all of the six winning numbers is part of the standard program of every introductory class in probability theory. One has to enumerate in how many different ways it is possible to pick exactly six of the numbers 1, 2, ..., 49. There are 13,983,836 possibilities. In the German 6/49 lottery, all of these possibilities are equally probable, and no expense or effort is spared to guarantee this uniform probability distribution. Thus, for each individual combination of six numbers, the probability of being drawn is exactly

$$1 : 13,983,836.$$

Table 12.1 Winning probabilities for a single play in the German 6/49 lottery. The play consists of choosing six of the numbers between 1 and 49. The distribution shares refer to the percentage of the handle available in a prize class after deducting fixed prices in class IX

	Number of correct choices	Distribution share	Chance 1 to
Prize class I	6 + bonus number	12.8%	139,838,160
Prize class II	6	10%	15,537,573
Prize class III	5 + bonus number	5%	542,008
Prize class IV	5	15%	60,223
Prize class V	4 + bonus number	5%	10,324
Prize class VI	4	10%	1147
Prize class VII	3 + bonus number	10%	567
Prize class VIII	3	45%	63
Prize class IX	2 + bonus number	Fixed prize	76

Adding the bonus number reduces this probability by a factor of 10. And for 6 correct numbers without the bonus number, the probability is given in Table 12.1. For the other classes, the calculation is a little more complicated, but far from a major mathematical feat.

12.2   Optimizing the Payout

Unlike the winning probabilities from Table 12.1, the prizes as such are to a considerable extent open to manipulation. As total prize money in each class is distributed among respective winners, the individual winner receives little if there are many winners. If there are few winners, the individual winner receives a lot.

On November 26, 2011, the individual winners received little—less than 30,000 euro for six correct numbers—no fewer than 78 players had picked the winning combination 3-13-23-33-38-49. The winners would have been even more disappointed if the numbers 1 to 6 had won—these are picked by more than 50,000 bettors every Saturday. Also the patterns in Fig. 12.1 were picked more than 3000 times on a given weekend in Baden-Württemberg alone and more than 20,000 times throughout Germany.

As can be seen, geometric patterns are very popular. The last combination is an exception—the winning numbers of the week before. If such popular patterns are actually drawn, payouts are very low. Figure 12.2 shows another winning example where just 53,000 Marks were paid out for 6 correct numbers.

Gamblers who pick such series are behaving unwisely. Smart players choose combinations that they have for themselves. There are a lot of these, as testified by the frequent occurrence of draws where all correct numbers were picked by nobody. A good strategy to find these unselected combinations is to imitate chance: put 49 pieces of paper into a hat and pick 6 of them at random. It would be beneficial to exclude all combinations in which geometric or arithmetic patterns occur. It is

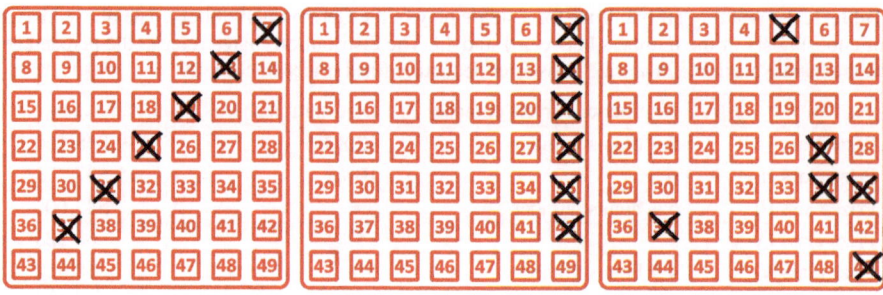

Fig. 12.1 Popular German lotto combinations

Fig. 12.2 The winning
numbers on October 4, 1997.
They were picked by 124
bettors

hard to believe how many people are proud of knowing the first six prime numbers, and then mark them on their lottery ticket. Additionally one should throw away all combinations involving a 19; this number is incredibly popular with bettors. In general, payouts are below average if the majority of the numbers drawn is below 30. Also, past winning combinations are to be avoided, like the one in Fig. 12.2. Even today, every weekend, several dozens of lottery gamblers pick the numbers of the first drawing on October 9, 1955. And indeed, it has happened that the same combination of six numbers has been drawn a second time: The 15-25-27-30-42-48 numbers of December 20, 1986 came up a second time in the drawing A of the Wednesday lottery on June 21, 1995.

So avoid these mistakes when selecting your numbers, proceed to the lottery retailer and hand your ticket in. Do not tell anyone about this system—it only succeeds if the majority of all bettors continues to produce conscious patterns of various types and leave sufficiently many unselected combinations left. If all bettors picked their numbers randomly, lottery tickets would be evenly distributed across all 139,838,160 possibilities, and the expected payout, i.e. exactly 50 cents for every euro wagered, would be the same for all.

12.3   Further Reading

Karl Bosch (1994): "Lotto und andere Zufälle - Wie man die Gewinnquoten erhöht," Oldenbourg Wissenschaftsverlag; John Haigh (1997): "The statistics of the national lottery," Journal of the Royal Statistical Society A, 160, 187–206; Norbert Henze and Hans Riedwyl (1998): "How to win more - Strategies for increasing a lottery win," Routledge.

Part III
Money and Business

Chapter 13
Statistics at the Stock Exchange

Walter Krämer and Tileman Conring

Abstract For decades, prices and returns of financial assets have been a coveted object of statistical research. The present chapter focuses on common stocks and presents some recent contributions out of the enormous literature which has been accumulated on this subject. Unfortunately statistics does not teach us how to become rich with stocks (except in the long run). But statistics can certainly help us not to become poor.

13.1   Beware of Dependencies

September 15, 2008 was a day of disaster. On this Monday the famous Lehman Brothers investment firm went into bankruptcy. About 20,000 Lehman employees lost their job immediately, and in the sequel many more employees, and those of other firms, also lost their jobs due to the bankruptcy. Economies worldwide went into recession and the year after, the German GDP fell by about 5%. This was so far the largest drop since World War 2.

Among the many causes of the Lehman-disaster—some consider it the most important one—was a serious but unrecognized defect in many structured financial products offered by Lehman and many other major banking institutions at that time: a considerable underestimation of dependencies among the various ingredients making up the structure of these products. If one component of a product defaults, what is the probability that the same happens to others? If this probability is small, there is little reason to worry—this resembles crossing the Atlantic in a four-engine jet plane. If one engine fails, you still get safely home. But if three engines fail, disaster looms.

Among the important engines of the financial successes preceding the Lehman crisis were housing credits. Clever financial engineers created allegedly almost risk-

W. Krämer (✉) · T. Conring
TU Dortmund, Department of Statistics, Dortmund, Germany
e-mail: walterk@statistik.tu-dortmund.de; tileman.conring@tu-dortmund.de

C. Weihs et al. (eds.), *Statistics Today*, Society, Environment and Statistics,
https://doi.org/10.1007/978-3-662-68907-3_13

free investments by disregarding the severity of mutual default dependencies. In reality, these 'risk-free' investments were full of risks. And when the housing bubble burst, even the most inexperienced newcomer to financial investing learned about the importance of these dependencies.

13.2   Investing in Stocks

Dependencies are also essential when investing in stocks. And in spite of occasional crises like the Lehman disaster, this is in the long run still the most profitable investing route. If at the beginning of 1958 grandpa Heinz had invested 10,000 German marks in stocks of the German stock price index DAX and had forgotten about this investment ever since, after 60 years the initial investment would have grown to be worth 200,000 euros, corresponding to an annual return of 7.8%. This is far above of what he could have earned via fixed investments of any type (see Fig. 13.1).

When investing in risky assets, whether stocks or bonds or real estate, the most important decision concerns the composition of the portfolio. That it is unwise to put all one's eggs into one basket has been known for a long time. But sophisticated statistical portfolio theory originated only in the late 1940s/early 1950s at the University of Chicago where a young PhD student named Harry Markowitz (later to earn the noble prize in economics for his discoveries) was completing his dissertation. Like everybody else, he focused upon the two most important criteria for any financial investment: expected return and risk. Expected return is what one earns in the long run. For stocks and in real terms this is about 5% per year. Actual returns, of course, exhibit considerable fluctuations and are often also negative. The average amount of such fluctuations is the conventional measure of risk. It has long been noted that investments with high long time returns also exhibit a larger amount

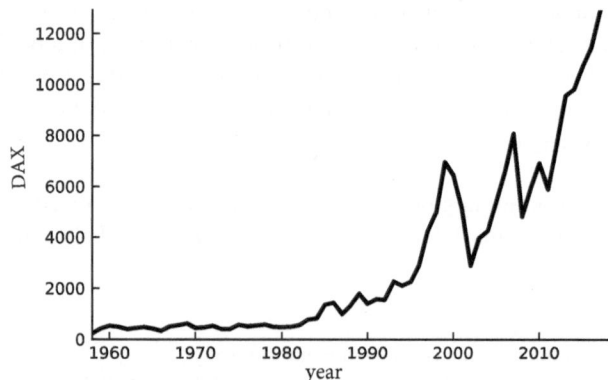

Fig. 13.1 Development of the DAX from 1958 to 2018

of risk at the same time. Markowitz, in his seminal contribution, was able to show how to build a portfolio such that, given some expected return, the level of risk is minimized. Alternatively, given some level of risk that the investor is prepared to tolerate, Markowitz shows how to find a portfolio such that the expected return is maximized. This is a very clever idea and the noble prize is certainly a very appropriate acknowledgement for this achievement.

13.3 Time-Varying Dependencies

The most important ingredient in the Markowitz formula is the dependence structure of the returns of the assets that make up the portfolio. Applying the Markowitz formula therefore requires knowing these dependencies. If these remain constant for some time they can easily be estimated from historical data using standard statistical textbook methods.

Unfortunately, dependencies can change. Table 13.1 shows annual returns of the world's major stock markets for two years: The year before the Lehman disaster (2007) and the disaster year (2008). The pre-crisis returns are as expected: on average positive, but not uniformly so. Some countries like China are booming, others, like Japan or Italy, are in mild distress.

The disaster year is different: all markets are in severe distress, or in technical terms: dependencies have dramatically increased. This is also known as the great 'diversification meltdown': exactly when we need it most, the diversification effect ceases to exist. Therefore, statisticians worldwide and at the Dortmund statistics department in particular are working on methods to model and forecast such changes in dependence.

Similar fluctuations show up in daily returns. Figure 13.2 plots daily returns of the German stock price index DAX and the American S&P500. The panel to the left plots returns prior to the crash, the panel to the right plots returns thereafter. As can be seen, both the variability and the correlation of returns have drastically increased. Before the crash daily returns never exceed 5% in absolute value, whereas

Table 13.1 The big 'diversification meltdown': Performance of global stock markets 2007 and 2008

Country	2007	2008
USA (DJIA)	+6.4%	−32.7%
Japan (Nikkei 225)	−11.1%	−29.5%
Germany (DAX)	+22.3%	−39.5%
GB (FTSE 100)	+3.8%	−30.9%
France (CAC40)	+1.3%	−42.0%
Spain (IBEX 35)	+7.3%	−38.7%
Italy (S+P Mib)	−7.0%	−48.8%
China (Shanghai Comp.)	+96.7%	−65.4%
India (Sensex 30)	+47.1%	−52.9%

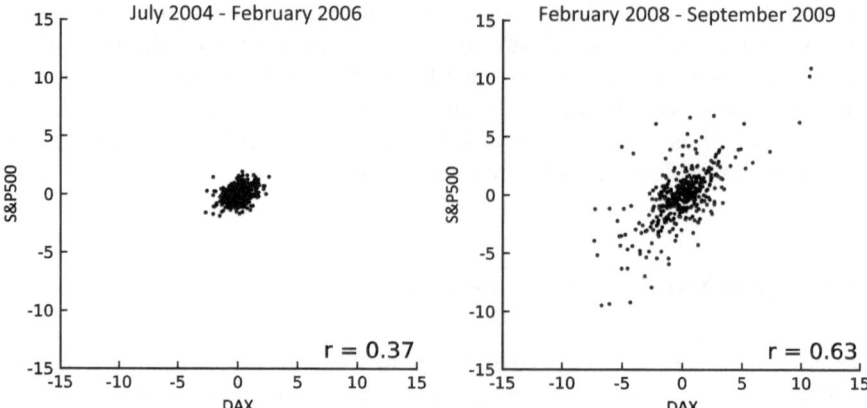

Fig. 13.2 Daily returns of the German stock price index DAX and of the American S&P500

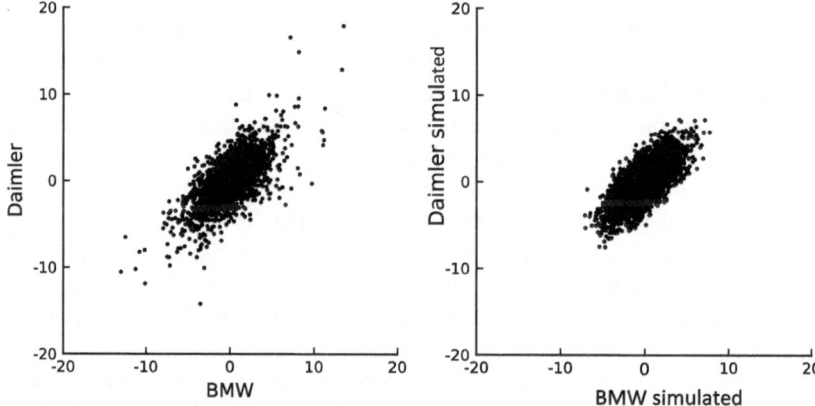

Fig. 13.3 Tail dependency of stock returns. Left: 4000 daily BMW and Daimler returns, Right: 4000 bivariate normally distributed random vectors

thereafter, this happens more than a dozen times (with—surprisingly—the largest returns in absolute terms being positive, both plus 11% on a single day in Germany and in the USA, both at October 10th, 2008). However, dependencies also increased from a pre-crisis correlation coefficient of 0.37 to a post-crisis coefficient of 0.63.

The phenomenon that particularly large returns, whether positive or negative, tend to happen simultaneously is referred to in statistics as tail dependence. Such tail dependencies seem to be common in stock returns. Figure 13.3 compares two bivariate distributions. The panel on the left shows 4000 daily returns of BMW and Daimler, the panel on the right shows a bivariate normal distribution with the same first and second moment as the panel to the left. Obviously, joined extremes occur much more often in real data as compared to an artificial distribution with the same lower order moments. Such tail dependencies are the focus of research in Dortmund and many other statistics departments throughout the world.

13.4   The Not So Normal Normal Distribution

Traditionally, returns of risky assets have been modeled as normally distributed random variables (the normal distribution is the one with the well-known bell curve as its density). For stock prices this distribution can not fit exactly if returns are defined in the usual way as

$$\frac{\text{new value} - \text{old value}}{\text{old value}}.$$

This return can never be smaller than $-1 = -100\%$. But normally distributed random variables can take on arbitrary small values (although with very small probabilities). For this and other reasons returns are often modeled as

$$\log\left(\frac{\text{new value}}{\text{old value}}\right).$$

As the mathematically inclined will easily verify, this is the return which transforms the old value into the new one in the case of continuous reinvestments. But even if defined like that, stock returns are normally distributed only approximately at best. Figure 13.4 shows a histogram of 11,000 daily returns of Deutsche Bank with the corresponding bell curve ('corresponding' means the same mean and variance).

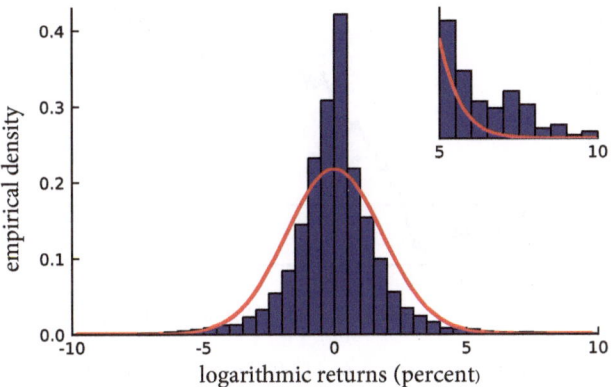

Fig. 13.4 Distribution of returns of the Deutsche Bank; top right: right margin enlarged

It is seen in Fig. 13.4 that the actual values in the center and at the tail of the curve occur more often than prescribed by the normal distribution. In professional jargon this is called excess kurtosis. In particular, the tails have provided much food for thought in capital market theory in recent decades. If tail probabilities tend to zero too slowly, the underlying distribution does not have any higher moments. In the extreme case of a Cauchy distribution, not even the expectation exists. However, this renders all of conventional portfolio theory useless, which relies on finite second moments. This problem has also been a subject of much research in recent decades.

On the other hand, little doubt has been cast on the underlying distribution's symmetry. Here we found a pattern which has so far been ignored. If one confines oneself to small returns, say less than 1% in absolute value, negatives are more frequent than positives. Plotting sums of small returns below some critical value against this critical value produces a graph like Fig. 13.5.

We call this the spoon effect. This spoon effect vanishes if returns of two or more stocks are averaged. It also does not exist for stock price indices. This points to a company-specific explanation. This matter is currently under investigation.

Alternatively, one might sort returns according to absolute value. Plotting such sorted returns produces a pattern similar to Fig. 13.5. In the beginning, but also following the maximum, the sum of ranked returns tends to decrease.

Figure 13.6 shows daily returns of Deutsche Bank starting 1973 up to 2015. The maximum of absolute returns according to absolute value occur at position 9927. The following 505 absolute returns show a negative tendency and let the overall return decrease to 111. So, 5% of trading days are sufficient to annihilate all the gains accumulated in the others. There is much research going on how to integrate this type of asymmetry into efficient pricing.

Fig. 13.5 Spoon curve: cumulated returns of 11,000 Deutsche Bank returns as a function of the absolute value of the maximum

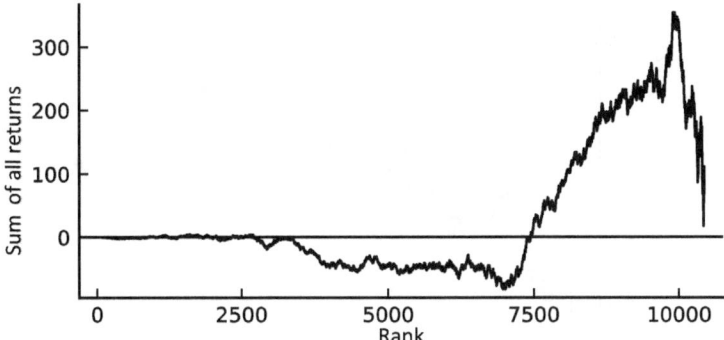

Fig. 13.6 Cumulated return in dependence of its rank

13.5 Cointegration

Logarithmic returns have many advantages. They can, better than standard returns, be modeled by the normal distribution (though not perfectly, as shown above). Even more importantly, weekly returns are the sums of daily returns. Assuming in addition zero correlation plus equal expectations and variances would then render the log stock price itself a perfect random walk with (a positive) drift, i.e. a sequence of random steps with a tendency to rise (in Fig. 13.5, e.g., for maximum absolut returns between 1.5% and 4%, say). In technical jargon: the log price is integrated of the order one (I(1)). A famous insight which has generated two noble prizes for its discoverers concerns the joint behavior of I(1) processes. When there is a linear combination which is no longer I(1), the initial processes are called cointegrated.

Cointegration implies that current changes can be forecasted from past values. This is impossible in efficient stock price markets where by definition current changes are independent from the past. Therefore cointegration of stock prices implies a contradiction of the efficient market theory and the possibility to make money by forecasting.

In practice, however, once in a while cointegration seems to happen. Figure 13.7 shows log prices of VW shares, both standard and preferred. According to all rules of mathematics and statistics, both price series follow an I(1) process. But the difference (which is a very simple type of linear combination) obviously does not. Therefore, changes in price can be forecasted from previous stock price values, and there have been lots of efforts in the past to do so. It seems, however, that the frequency of transactions required by such strategies generates costs which outweigh any gains.

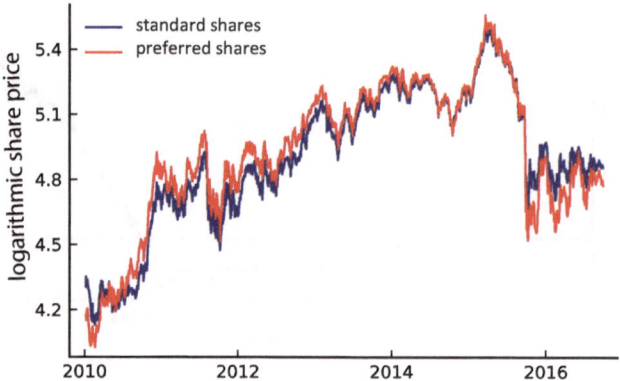

Fig. 13.7 Price development of the VW standard and preferred shares

13.6 Further Reading

The topics of this chapter were intensively studied in a collaborative research center at the department of Statistics of the TU Dortmund University. The Nobel Prize awarded findings of Clive Granger and Robert Engle about cointegration appeared in "Cointegration and error correction," Econometrica 1987, 251–276. Our own theories on cointegration and the statistical properties of German stock returns are, inter alia, published in Krämer, Runde (1996): "Stochastic Properties of German stock returns," Empirical Economics 21, 281–306, and Krämer (1999): "Kointegration von Aktienkursen," Zeitschrift für betriebswirtschaftliche Forschung 51, 915–936. On the prevalence of negative and high absolute fluctuations consider Lempérière et al. (2017): "Risk premia: asymmetric tail risks and excess returns," Quantitative Finance 17, 1–14.

Chapter 14
Statistics in the Risk Assessment of Bank Portfolios

Dominik Wied and Robert Löser

Abstract Better safe than sorry. This also applies to risky investments on the stock market. And this is especially true for those who speculate with other people's money, i.e. banks. In recent years, statistics has provided ways to gain control of such risky investments.

14.1 The Problem

Worldwide, shares issued by German companies are currently worth 1.8 trillion euros (as of June 2017). Naturally, it is of great relevance to their owners how much their assets are still worth at the close of the market on the next day, or a week, a year, etc. later. Furthermore, the probability of loss in the development of the shares is of great interest to the shareholders. For banks, the estimation of this loss is even prescribed by the government in the so-called Basel III guidelines. Among other things, these guidelines determine the amount of equity to be held, which is intended to prevent banks from becoming insolvent, as happened hundreds of times during the so-called Lehman crisis. The causes of this financial crisis have been the subject of much discussion in its aftermath; one explanation repeatedly put forward relates to incorrect risk assessment. According to this explanation, many banks had underestimated the risks of their portfolios and, as a result, held too little equity capital, so that they were no longer able to service the suddenly higher liabilities. The correct estimation of the highest possible disaster is a prime example of how applied statistics can serve risk assessment.

D. Wied (✉)
University of Cologne, Institute for Statistics and Econometrics, Köln, Germany
e-mail: dwied@uni-koeln.de

R. Löser
TU Dortmund, Dortmund, Germany
e-mail: robert.loeser@tu-dortmund.de

14.2   Expected Shortfall Compared to Value-at-Risk

In the following, X indicates the change in value of a risky asset up to a specified deadline, for example until the close of trading on the next day. Since we are mainly interested in potential losses, we define X as

$$X = \text{current value} - \text{future value}.$$

A positive X therefore represents a loss, a negative X corresponds to a gain. That is why X is also called risk. To assess this risk, we are in search for a formula which assigns a so-called risk measure $\varphi(X)$ to this random change X. According to up-to-date and widely accepted capital market research, this rule should meet the following minimum requirements:

- Translational Invariance:
  If one adds a secure risk of X to a risk c, the risk measure increases by exactly c:

$$\varphi(X + c) = \varphi(X) + c$$

- Positive homogeneity:
  If a risk is increased/decreased by the positive factor λ, the risk measure also increases/decreases by exactly the factor λ:

$$\varphi(\lambda \cdot X) = \lambda \cdot \varphi(X)$$

- Monotony:
  For two risks X and Y where Y under all circumstances realizes a loss at least as big as X, the risk measure of X is less than or equal to that of Y:

$$X \leq Y \ \Rightarrow \ \varphi(X) \leq \varphi(Y)$$

- Subadditivity:
  The risk measure of a portfolio which entails the two risks X and Y is less than or equal to the sum of the individual risk measures. This corresponds to the intuition that one can generally reduce risk through the diversification of stock purchases:

$$\varphi(X + Y) \leq \varphi(X) + \varphi(Y)$$

A risk measure that satisfies all four requirements is called 'coherent'.

In practice, two main risk measures are used. One is the value-at-risk VaR (at a level α), formally defined as the $(1-\alpha)$-quantile of the loss distribution (i.e., the value that is exceeded with a probability of exactly α). The other is the Expected Shortfall ES (at a level α), which is defined as the loss which is expected if the $\alpha\%$-VaR is exceeded. In accordance with common values of significance levels, α is typically set to 5, 2.5, and 1.

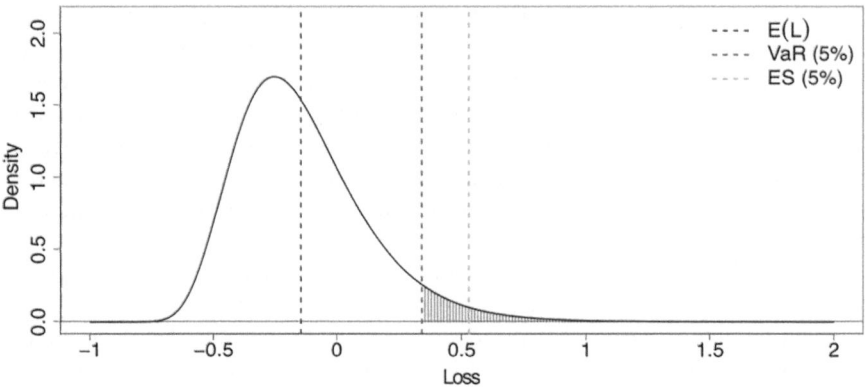

Fig. 14.1 VaR and ES of loss L. The expected value E(L) of the loss is negative, i.e. we expect a gain

In general, the higher the VaR or ES, the more equity the bank must deposit to hedge the risk. Figure 14.1 illustrates these two measures and the total expected loss E(L) in a hypothetical loss distribution. How this loss distribution can be determined is open to question. In practice, this constitutes a big problem that cannot be dealt with here.

Unfortunately, the VaR does not indicate the exact size of the loss, if it is exceeded (cf. the hatched part of the curve right to the red line). In practice, however, it may well be relevant whether 5% or 20% of the portfolio value is lost on average. This information is of increasing interest in the financial industry and is also a requirement in recent regulations on capital adequacy, as specified by the financial supervisory authorities. For this reason, the ES has started to replace the VaR as the standard measure of risk. Moreover, the ES comes with the statistical advantage of being coherent. In contrast, the VaR does not generally satisfy the axiom of subadditivity; it may happen that diversification of a portfolio increases the VaR. The ES, on the other hand, is more difficult to estimate and also more difficult to validate. Therefore, the VaR is still the dominant measure.

14.3   Estimation of Risk Measures

In order to work with risk measures, they must first be estimated on the basis of suitable historical data. A distinction must be drawn between nonparametric methods, which are based exclusively on observed losses, and parametric methods, which are model-based. The simplest non-parametric method for VaR estimation is the historical VaR, i.e. the empirical quantile of observed values. For example, if the risk manager were to look at the 1000 most recent data points (i.e., approximately four years on a daily basis), he or she would estimate the 5%-VaR by sorting the loss values in ascending order and using the value at the 950th position as the estimate. A more complicated procedure is the historical simulation which is carried out by means of a so-called bootstrap procedure. In this approach, a new time series of the same length as the historical one is artificially generated from past data using 'sampling with replacement' from the historical loss values. In a next step, the empirical quantile of the simulated data is determined. This is repeated several times and, finally, the mean value of the empirical quantiles serves as an estimate of the true quantile.

The described nonparametric methods are, in principle, also applicable for estimating the ES and work well if the data basis is sufficiently large, e.g. if a sufficient number of years of daily losses is available. For intraday data, the time period may be shorter. If the data set is too small, parametric methods are better suited for estimation. This applies in particular to the estimation of the ES for which, in the case of non-parametric methods, effectively only $\alpha\%$ of the data can be used. A fairly simple parametric method is to fit a normal distribution to the losses whose two parameters (expected value and variance) are assumed to be constant over time. The risk manager then estimates these parameters using the empirical mean and variance and then takes the risk measures from the fitted loss distribution.

In simple estimation methods, certain statistical properties of the risks that deviate from the norm can lead to erroneous estimates and to too low risk measures. Typical deviations are heavy margins (i.e. extreme events occur relatively often), marginal dependencies (e.g. high losses often occur together), or time-varying market parameters (e.g. calm and volatile market phases). We have developed solutions which can deal with such properties. For example, in the case of 'heavy margins', it has been shown that larger losses occur more frequently than can be explained by a normal distribution (cf. Chap. 13). This is a problem especially for ES estimation. Other distributions, e.g. the t-distribution or special extreme value distributions, offer alternatives to solve the problem of heavy margins.

The assumption of constant variance is also questionable, e.g., for the case of time-varying market parameters. Already decades ago, empirical evidence showed that risk distributions tend to form so-called volatility clusters, as Fig. 14.2 illustrates using historical DAX (percentage negative) risks. Days characterized by high swings are often followed by further days of high swings. The classical approach to this phenomenon is risk forecasting using so-called ARCH and GARCH models. Here, current variance is modeled as a function of past variances and risks.

14.4   Validation of Risk Models

Estimates for VaR or ES must be validated before use. Banks that use their own models for estimation, which in turn determine the amount of equity to be deposited, are even obliged to do so. At present, the required equity is determined using the 1%-VaR. If a bank uses a model for VaR estimation, it must be validated over a period of 250 days. During the validation period, the 1%-VaR for the next trading

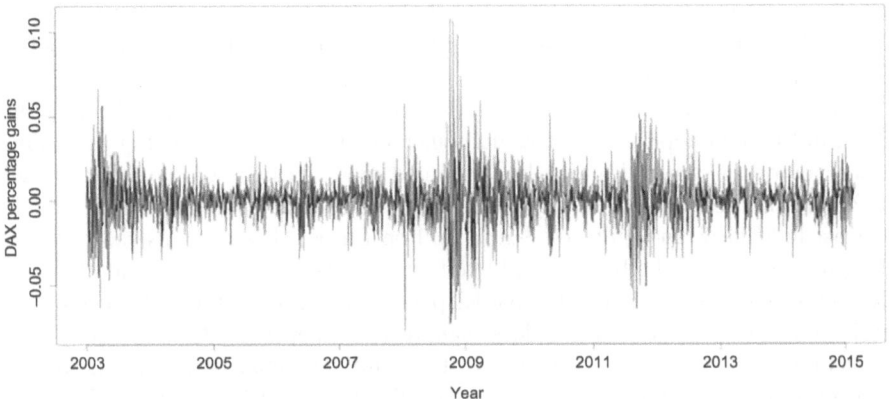

Fig. 14.2 Daily percentage gains (= percentage negative risks) of the German stock index (DAX)

Table 14.1 Basel traffic light

Zone	Number of VaR exceedances	Factor
GREEN	0	0.00
	1	0.00
	2	0.00
	3	0.00
	4	0.00
YELLOW	5	0.40
	6	0.50
	7	0.65
	8	0.75
	9	0.85
RED	≥ 10	1.00

day is estimated and then checked to see whether the actual loss has exceeded the VaR. If this is the case, the corresponding day is coded with by the value 1, if not, by means of 0. After 250 days, this coding has produced a series of 0s and 1s. For an ideal model, one would expect $250 \cdot 0.01 = 2.5$ VaR exceedances. As soon as exceedances occur significantly more often, one has to question the model used or even reject it. The Basel Committee on Banking Supervision prescribes a 'traffic light approach' (Table 14.1) to assess the severity of exceedances.

If the model used for estimation (also called risk model) falls into the category RED, it must be improved immediately or replaced by another model. The probability of a correct model mistakenly falling into this category is less than 0.1%. In Table 14.1, the value 'Factor' influences the equity that needs to be held back as follows:

$$\text{Equity held} \geq (3 + \text{Factor}) \cdot \text{VaR}.$$

The formula indicates that the equity capital of banks must be high enough to service the 1%-VaR at least 3 times at any given time; in the case of a doubtful model, it needs to be even higher.

This approach has long been criticized (as has the VaR itself), in part because the Basel traffic light approach accepts dubious models too generously. Let us assume a bank carries a standard normally distributed risk over one year with a true VaR of 2.33 million Euros. According to the formula introduced above, the bank should at least hold back an equity of 3·2.33 million Euros. However, instead of using this prescribed VaR, the bank would like to reduce its equity. To this end, the bank decides to be satisfied if its risk model passes the Basel traffic light test with a probability of 90%, i.e. if it has less than 10 exceedances in 250 days with 90% probability. Under these circumstances, the bank can actually declare its VaR at 1.96 million. This would result in a probability for GREEN of approx. 25%, for YELLOW of 65%, and for RED of 10%. The 90% criterion would therefore be fulfilled and the bank could reduce its required equity. However, this conclusion does not always hold, since in 65% of the cases (yellow zone), the Factor is greater than zero and this, from a certain value onwards (here 0.65), leads to an equity even higher than that for the true VaR. To prohibit such illicit manipulations by the banks, we need better-suited statistical models than the Basel traffic light and even stricter regulations. Therefore, validation procedures, so-called backtests, have been proposed in mathematical statistics, which identify incorrect models with higher probability.

Furthermore, the Basel traffic light approach only considers the pure number of exceedances instead of taking into consideration their overall distribution. In a good model, such a distribution should not show any clusters of exceedances to maintain the assumption of independence of exceedances, as illustrated in Fig. 14.3, timeline (a). In (b), on the other hand, we see an unfavorable case, in which the VaR exceedances are very close to each other, i.e. a series of unexpectedly high losses

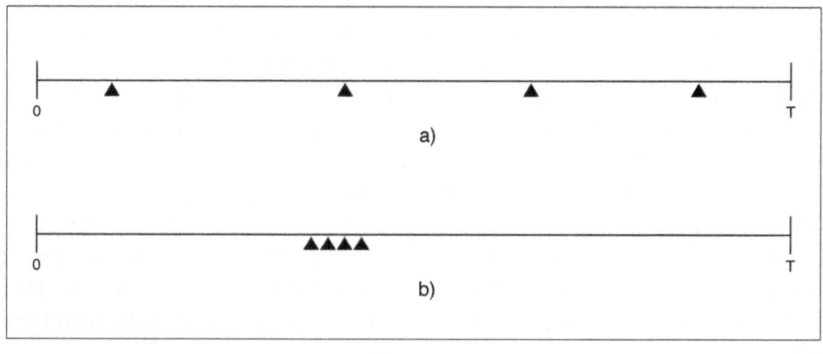

Time

Fig. 14.3 Different distributions of VaR-exceedances over time. (**a**) No clusters in the VaR exceedances. (**b**) Clusters in the VaR exceedances

occurred in a short time period. In an ideal model, such a case would not occur (with high probability). This is because the probability of a VaR exceedance in this model is always $\alpha\%$, regardless of whether or not there was a VaR exceedance on the previous day. Consequently, no dependencies should be found in the sequence of 0s and 1s (independence property; see above). To test for independence, we have developed various statistical tests to detect possible deviations.

As argued above, in a good VaR model, the distances between exceedances are independent of each other. In the case of ES, the approach is more complicated. This is also one of the main reasons why VaR is currently the predominant approach. If we want to investigate the distribution of exceedances in relation to the ES, one can look at all losses greater than VaR and divide them by the estimated ES:

$$\frac{\text{loss when VaR is exceeded}}{\text{estimated ES}}.$$

From these ratios, we determine the mean. This mean should, per definition, be close to 1. The risk model is discarded if the mean value is significantly greater than 1.

An alternative approach we developed evaluates VaR exceedances on a scale from 0 to 1. If exceedances are high, we assign values close to 1; if they are close to VaR, they are assigned a value close to 0. In an ideal model, these values should be randomly realized somewhere between 0 and 1, and should be 0.5 on average. If the mean of these exceedances is significantly different from 0.5, the model used should be questioned or replaced by another model. For a mean value smaller than 0.5, the model would overestimate the risks, for a value larger than 0.5, it would underestimate them. For the problem under investigation, it is particularly important to detect the latter case in order to avoid underestimating risks in the future.

14.5   Further Reading

The reference work of modern risk measurement is McNeil, Frey, and Embrechts (2009): "Quantitative Risk Management: Concepts, Techniques and Tools," Princeton University Press. Our own suggestions for improved backtests can be found in D. Ziggel, T. Berens, G. Weiß, and D. Wied (2014): "A New Set of Improved Value-at-Risk Backtests," Journal of Banking and Finance 48, 29–41, R. Löser, D. Wied, and D. Ziggel (2018): "New Backtests for Unconditional Coverage of the Expected Shortfall," Journal of Risk 21(4), 1–21.

Chapter 15
On Rating the Raters: Statistics in the Rating Industry

Walter Krämer and Simon Neumärker

Abstract Individuals and enterprises have never been as indebted in the history of mankind as they are today. Net indebtedness is of course always zero, as one party's obligations are always another party's assets. Nevertheless, the desire to secure repayment and to know the risk of default were probably never greater than they are today. This has created a huge market for suppliers of default predictions, in conjunction with a need to judge their forecasting capabilities.

15.1 Obligations and Obligors

> There are two superpowers in the world today in my opinion. There's the United States and there's Moody's Bond Rating Service. The United States can destroy you by dropping bombs, and Moody's can destroy you by downgrading your bonds. And believe me, it's not clear sometimes who's more powerful.

This quote is from 1996; it is from an interview with the New York Times journalist Thomas Friedman. Although certainly a bit exaggerated, it hits a nerve. There are a lot of government officials worldwide who do not dread anything more than being downgraded by any of the major rating agencies. At the time of writing this chapter, federal and local German governments have accumulated public debts of about 2000 billion euros. The costs of servicing this debt vary by the date of issue. Currently, lenders will loan money to Germany at no interest. But in the long term, public debtors with an optimal AAA rating pay about 3% interest per year, which in the case of Germany amounts to an annual debt service of 60 billion euros. If Germany were rated only A, it would have to pay 6% which is another 60 billion euros on top. This is certainly worth thinking about.

W. Krämer (✉) · S. Neumärker
TU Dortmund, Department of Statistics, Dortmund, Germany
e-mail: walterk@statistik.tu-dortmund.de; simon.neumaerker@tu-dortmund.de

Table 15.1 Public indebtedness of selected countries

Country	Indebtedness (in billion €)	Per capita (in €)	In % GDP	Rating Moody's	S&P	Fitch
USA	19,617	58,442	108	Aaa	AA+	AAA
Japan	11,038	87,772	266	A1	A+	A
France	2638	39,315	114	Aa2	AA	AA
Italy	2531	42,012	149	Baa3	BBB	BBB−
Germany	2278	27,391	67	Aaa	AAA	AAA
Greece	334	31,187	187	Ba3	BB−	BB
Sweden	176	17,042	37	Aaa	AAA	AAA

Table 15.1 presents the current public indebtedness of selected countries, both in absolute terms and relative to gross national product, in conjunction with its rating by the three leading agencies: Moody's, Fitch, and S&P.

As can be seen, the United States have accumulated the largest debt; however, this is only average in the perspective of debt as a percentage of GDB. Things are dramatically different in Japan, but there is obviously little doubt that Japan will be able to repay its debts (even a single A is still considered good investment grade). In the case of Greece and Italy, one is not so sure. The coveted AAA is reserved for Germany and Sweden, which implies that the German Federal Ministry of Finance currently does not have to pay any interest for new debts anymore. It even gets some money back which has never happened in the history of humanity before.

AAA means there is no risk of default even in the long run. Besides Germany and Sweden, only Luxemburg, Australia, Denmark, Norway, the Netherlands, Switzerland, and Singapore can currently claim this rating.

Awarding such ratings is a lucrative business and the respective companies charge heavily for their services. Table 15.2 shows the firms which share this market in the EU. The leader in terms of market share is the American Standard & Poor's Corporation (S&P) which came into being in 1941 by a merger of the H. W. Poor Corporation and the Standard Statistics Bureau. Its largest competitor is Moody's Corporation, founded in 1909 by John Moody, who at that time started rating railway corporations. Each of these firms bestows about one million ratings per year. The third major competitor is the New York and London based Fitch Ratings agency founded on Christmas Eve 1913 by John Knowles Fitch. Compared to these giants, the rest of the rating agencies are almost negligible in Europe. For instance, the European Rating Agency seen in Table 15.2 is located ironically in Slovakia and has to content itself with a very minuscule portion of the market left over by its big competitors.

Table 15.2 Market shares of rating agencies registered in the EU (European Securities and Markets Authority, 2017)

Rating agency	Market share in %
S&P Global Ratings	46.26
Moody's Investors Service	31.27
Fitch Ratings	15.65
DBRS Ratings	1.87
CERVED Rating Agency	0.97
AM Best Europe Rating Services	0.90
The Economist Intelligence Unit	0.69
CreditReform Rating	0.53
Scope Ratings	0.46
GBB-Rating	0.35
Assekurata	0.23
Euler Hermes Rating	0.22
Capital Intelligence Ratings	0.13
ICAP	0.12
ModeFinance	0.08
Spread Research	0.07
Dagong Europe Credit Rating	0.07
ARC Ratings	0.05
Axesor Rating	0.03
CRIF Ratings	0.03
BCRA Credit Rating Agency	0.02
EuroRating	0.01
INC Rating	<0.01
European Rating Agency	<0.01
Rating-Agentur Expert RA GmbH	<0.01
Total	100

15.2   How to Judge the Quality of Default Forecasts?

Not surprisingly, the firms listed in Table 15.2 do not let us know how they actually obtain their ratings. However, all ratings are based either implicitly or explicitly on forecasts of default probabilities. These probabilities are modeled as functions of selected explanatory variables with the coefficients of such regressions being estimated from historical data. Therefore, letter ratings in some way or the other correspond to numerical default probabilities. Although some agencies resent this view (arguing that such default probabilities vary over time), at the end of the day and at a particular point in time, an S&P A rating is equivalent to the statement: The probability that this client will default within one year is equal to—say—0.09%. It is therefore unfair to shame a rater if a client with such a high rating actually defaults. This can happen with a certain probability and, therefore, does happen once in a while. The most famous example is the Lehman Brothers bank which was rated A

Table 15.3 Estimated one-year default probabilities by S&P for corporate issuers (enterprises)

Rating	Default probability (in %)
AAA	0.003
AA+	0.006
AA	0.012
AA−	0.025
A+	0.047
A	0.091
A−	0.173
BBB+	0.299
BBB	0.495
BBB−	0.797
BB+	1.138
BB	1.518
BB−	2.280
B+	3.943
B	7.999
B−	19.557
CCC-C	48.355

almost until its famous default. Table 15.3 translates other letter ratings by S&P into one-year default probabilities.

Such probability forecasts are popular in other contexts as well. They originated in meteorology where statements like "the probability of rain in Berlin tomorrow is 5%" have been common for decades. The problem with such probability forecasts is that it is hard to judge their quality. What if it rains on one day out of 20 in Berlin in the long run? Then the forecast of a 5% rain probability is obviously correct, but it is also useless.

Probability forecasts are called 'well calibrated' whenever the long-run frequency of the forecasted event corresponds to the forecasted probability. But as the example above shows, calibration by itself is only a necessary but not a sufficient condition for a useful probability forecast. What one would desire, in addition, are forecasts which are spread towards the extremes of 0% and 100% probability. Extreme forecasts which are also calibrated are the optimum that can be possibly be obtained: they give a forecasted probability of 100% if the event actually occurs and 0% if it does not occur, a perfect foresight so to speak. Our challenge is to determine how close a forecaster gets to this optimum.

15.3   A Numerical Example

Assume that we have 800 obligors, out of which 160 default. This is 20%. We also have five raters A, B, C, D, E who attach default probabilities to each obligor (see Table 15.4). Most modest among these is rater A: he attaches a default probability label of 20% to all obligors. This is as trivial as it can get but it is still well calibrated. At the other extreme is rater E. He or she is omniscient and sorts all obligors exactly into defaults and non-defaults. And in between there are raters B, C, and D.

In this example, all raters are well calibrated. Obviously, A is the worst, and E is the best. But what about B, C, and D? Intuitively one would prefer raters whose ratings are more spread out. This criterion is also known as 'refinement'. In the example above, B is more refined than A and both C and D are more refined than B. Most refined of course is E. But what about the refinement relationship between C and D?

In mathematical jargon, a rater is more refined than a competitor if the competitor's forecast can be derived from the initial rater's forecast. Take C in the above example, and provide all 5% forecasts and a randomly selected half of all 15% forecasts with the label 10%. The rest receive the label 30%. This provides a well calibrated default forecast with the same stochastic properties as B's. Therefore, C is more refined than B.

However, C and D cannot be compared that way. Neither can D's forecasts be derived from C's nor C's from D's. Therefore, refinement provides only a partial ordering among well calibrated probability forecasts; there are forecasts which cannot be ordered according to this criterion.

Table 15.4 Distribution of default probabilities of five well calibrated probability forecasters

Forecasted default probability	Distribution of the credits among the default classes				
	A	B	C	D	E
0%	0	0	0	0	640
5%	0	0	200	80	0
10%	0	400	0	480	0
15%	0	0	400	0	0
20%	800	0	0	0	0
30%	0	400	0	0	0
45%	0	0	200	240	0
100%	0	0	0	0	160

15.4   Partial Orderings of Probability Forecasts

Apart and independently from refinement and calibration, it certainly makes sense to ask, in comparing A and B: Which of the two attaches higher default probabilities to actual defaults? This leads to the concept of default ordering. Some rater A dominates competitor B in this sense if he or she rates all defaulted obligors systematically worse in the sense that cumulated defaults starting with the worst rated obligors are always above those of B. However, this again provides only a partial ordering because these curves sometimes intersect. A related criterion is based on the receiver operating curve (ROC curve). For illustration consider rater D from Table 15.4. For 80 obligors he or she forecasts a default probability of 5% corresponding to a good rating. 480 obligors are assigned a default probability of 10% corresponding to a medium rating and 240 are assigned a default probability of 45% corresponding to a bad rating. Assume in addition that D is well calibrated. Among good ratings 4 still default which is 5% out of 80. In the medium group 48 obligors default which is 10% out of 480. In the third group 108 of obligors default which is 45% out of 240. Regrouping obligors according to their ratings starting with the bad ones and assigning them cumulated percentages of defaults leads to Table 15.5.

The worst group comprises 132 of the total of 640 non-defaults which is 20.63%. It also contains 67.5% of all defaults. The two worst groups together contain 564 of the total of 640 non-defaults which is 88.13%, and 97.5% of all defaults. Figure 15.1 gives the resulting ROC curve. Rater A is then better than B in this sense if A's ROC curve is nowhere below that of B.

15.5   Scalar Valued Measures of Forecasting Quality

The ROC curve is robust to monotone transformations of forecasted default probabilities. If all forecasted default probabilities are doubled, the forecast is no longer calibrated, but the ROC curve remains the same. Also, a rater who assigns identical default probabilities to all obligors has a ROC curve equal to the diagonal. Therefore, a ROC curve signifies a higher quality the more it is bent away from the diagonal towards the upper left corner of the figure. The area under the curve

Table 15.5 Defaults vs non-defaults in the rating classes for forecaster D

Rating	Cumulated percentages of total non-defaults	Cumulated percentages of total defaults
Bad	132/640 = 20.63%	108/160 = 67.5%
Medium	564/640 = 88.13%	156/160 = 97.5 %
Good	640/640 = 100%	160/160 = 100%

Fig. 15.1 ROC curve of
forecaster D: The more
defaults are in bad rating
classes, the more the ROC
curve bends away from the
diagonal

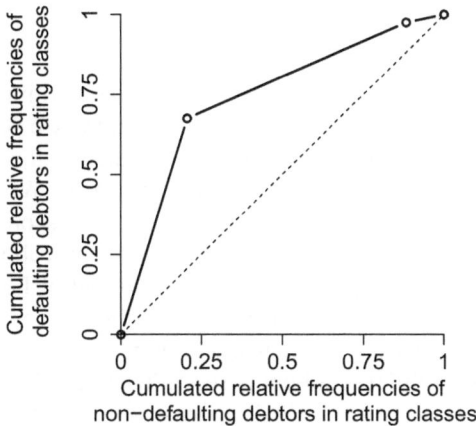

(AUROC) is often used as a quality criterion. Alternatively, one could rely on the maximal vertical distance from the diagonal.

In addition, there are lots of criteria which are based directly on the forecasted default probabilities. Most popular among them is the Brier score. Let p^j be the forecasted default probability for obligor number j (out of a total of n obligors to be rated overall) and set $\theta^j = 1$ for defaults and $\theta^j = 0$ for non-defaults. Then the Brier score is defined as

$$B = \frac{1}{n} \sum_{j=1}^{n} (p^j - \theta^j)^2.$$

This is simply the mean squared deviation of forecasts from realization (which are either 0 for non-default or 1 for default). Some refer to the negative of this score as the Brier score. It has traditionally been used to compare whether forecasts but is of course applicable to all contexts where probability forecasts are to be compared. It is confined to the (0,1) interval, the smaller, the better the forecast. The optimal value of 0 is obtained for forecasts which are either 0 or 1 and which are always correct. The worst possible value of 1 occurs for forecasts of exclusively 0% or 100% default probability where always the opposite of the forecast occurs.

There are many other suggestions for related scores which however introduce new problems of their own. For instance, a given pair of forecasters can contradict each other. Here we were able to show how to rule out such contradictions for incentive preserving scores. A score is called incentive preserving if it rewards honest forecasters: the subjectively expected score is maximized if one always uses one's personal beliefs for forecasting. The Brier score, for instance, is incentive preserving in this sense. We were able to show that a well calibrated forecaster B dominates another well calibrated forecaster A according to all incentive preserving score functions if its forecasts are more refined.

15.6 Further Reading

The valuation of probability forecasts originated with the meteorologist G.W. Brier. For a survey of the earlier literature see Krämer and Bücker (2011): "Probleme des Qualitätsvergleichs von Kreditausfallprognosen" (Problems of comparing the quality of credit default forecasts), AStA Wirtschafts- und Sozialstatistisches Archiv 5(1), 39–58. The relationship between score functions and refinement is discussed in, e.g., Krämer (2006): "Evaluating probability forecasts in terms of refinement and strictly proper scoring rules", Journal of Forecasting 25, 223–226, and in Krämer and Neumärker (2016): "Comparing the accuracy of default predictions in the rating industry for different sets of obligors", Economics Letters 145, 48–51. For the same with different obligors see Krämer (2017): "On assessing the relative performance of default prediction", Journal of Forecasting 36(7), 854–858.

Chapter 16
Gross Domestic Product, Greenhouse Gas Emissions, and Global Warming

Martin Wagner, Fabian Knorre, and Christina Kopetzky

Abstract Since the onset of the Industrial Revolution, the global mean temperature has risen by about 1 °C. It is beyond doubt that, amongst other factors, this increase has been driven by products of human activity, i.e. by carbon dioxide and other greenhouse gas emissions. What are the relations between economic activity and emissions? Do emissions necessarily increase with increasing economic activity?

16.1 Economic Activity and Emissions

How do emissions of the most important greenhouse gas, carbon dioxide (CO_2), which primarily emanates from the combustion of fossil fuels, depend on a country's economic activity? The term 'greenhouse gas' (GHG) stems from the fact that increasing concentrations of CO_2 and other greenhouse gases in the atmosphere cause rising temperatures. This is mostly due to reduced heat radiation from the earth's surface and the atmosphere into outer space. Figure 16.1 shows annual GHG emissions—including methane (CH_4), nitrous oxide (N_2O), and hydrofluorocarbons (HFCs) in addition to carbon dioxide—for Germany from 1990 to 2014. The trend for Germany is encouraging. Gross domestic product (GDP) increased by 41% in this period while GHG emissions expressed in CO_2 equivalents fell by 28%. Conversion to so-called CO_2 equivalents allows for analyses using a single

M. Wagner (✉)
University of Klagenfurt, Department of Economics, Klagenfurt, Austria

Bank of Slovenia, Ljubljana, Slovenia
e-mail: martin.wagner@aau.at

F. Knorre
TU Dortmund, Department of Statistics, Dortmund, Germany
e-mail: knorre@statistik.tu-dortmund.de

C. Kopetzky
University of Klagenfurt, Department of Economics, Klagenfurt, Austria
e-mail: christina.kopetzky@aau.at

C. Weihs et al. (eds.), *Statistics Today*, Society, Environment and Statistics, https://doi.org/10.1007/978-3-662-68907-3_16

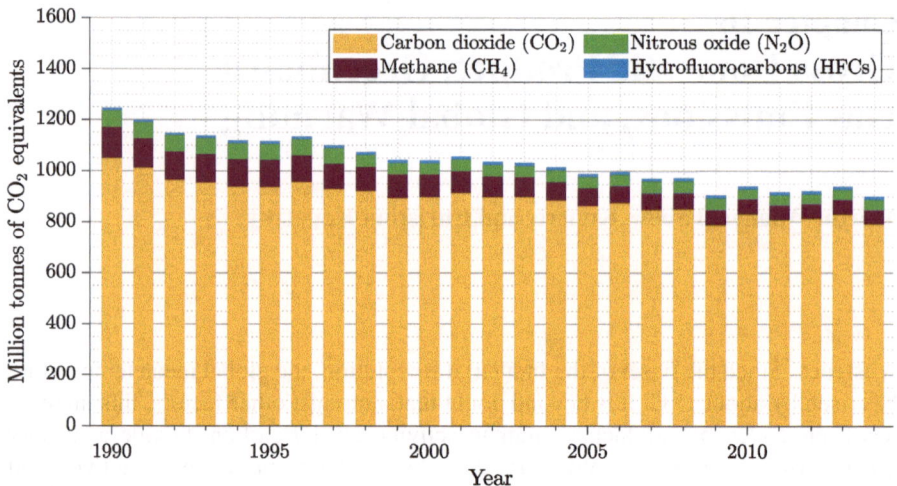

Fig. 16.1 Annual greenhouse gas emissions in Germany over the period 1990–2014 in million tonnes of CO_2 equivalents

aggregate value of emissions covering all gases. For instance, the conversion factor for methane is 25, which means that over a period of 100 years one tonne of methane causes the same GHG effect as 25 tonnes of carbon dioxide. Figure 16.1 furthermore shows that the vast majority of GHG emissions is caused by CO_2, for example, 88% in Germany in 2014.

However, the downward trend of Fig. 16.1 is a rather recent development, as can be seen in Fig. 16.2, which shows data from the early days of the Industrial Revolution around 1870 until 2014. It shows that emissions—expressed in logarithmic per-capita terms—in the four countries considered, i.e., France, Germany, the United Kingdom, and the United States, did not start decreasing until the 1970s. GDP per capita has been increasing since World War II, notwithstanding fluctuations due to recessions and booms. This means that a potential relationship between CO_2 emissions per capita and GDP per capita must be nonlinear. Following a period in which emissions and GDP both increased more or less strongly, some decoupling seems to have taken place as from the 1970s with GDP per capita growing continuously and emissions per capita decreasing. This trend can be observed not only for the four countries displayed in Fig. 16.2 but also for a number of other developed countries and even some less-developed countries.

Unfortunately, decreasing emissions per capita do not necessarily translate into a decrease in overall emissions. This is only true when population growth is slower than the decline in emissions per capita. Germany, for instance, had a population growth of about 3.5% from 1990 to 2014. This means that the emissions per capita actually even decreased by approximately 32% over the same period.

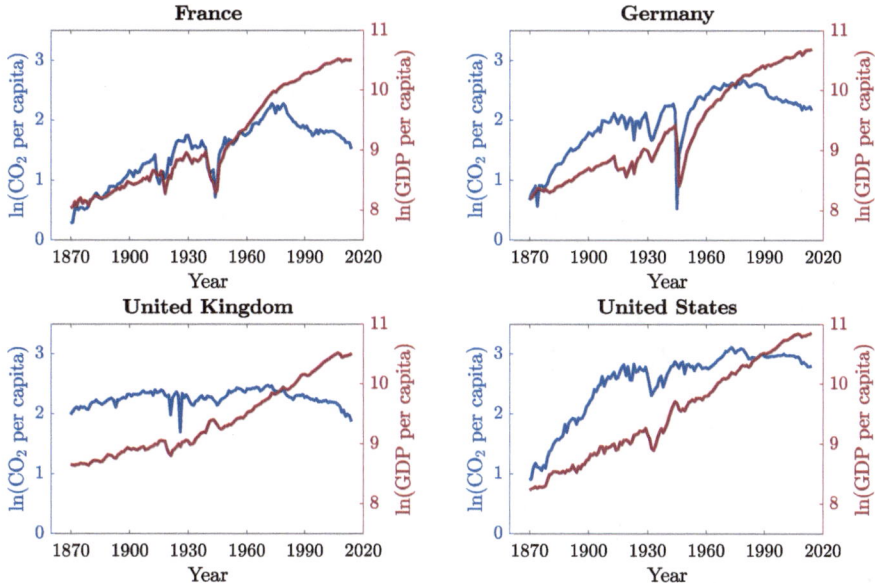

Fig. 16.2 Natural logarithm of annual GDP per capita and CO_2 emissions per capita over the period 1870–2014

The environmental impact discussed here for CO_2 emissions basically depends on three factors. First, on aggregate economic activity measured by gross domestic product: This is the scale effect. The scale of economic activity largely depends on the size and growth of the population. In order to adjust for the scale effect, which is also usually modeled separately in statistical analyses, the analysis is typically undertaken in per-capita terms like the per-capita quantities shown in Fig. 16.2. Second, on the sectoral composition of output: For a given technology, different goods will cause different emission intensities, i.e., kg of CO_2 emissions per euro or dollar of output. This is the composition effect. Third, on the technological effect: The emissions intensity generally decreases over time for various reasons, frequently due to technological changes and innovation, which often are consequences of changes in laws or regulations.

The most well-known hypothesis in the relevant literature is the so-called environmental Kuznets curve (EKC) hypothesis which postulates an inverted U-shaped relationship between the level of economic activity on the one hand and the environmental impact on the other hand, as shown schematically in Fig. 16.3. The interaction between the three factors determines the existence and shape of a potential environmental Kuznets curve.

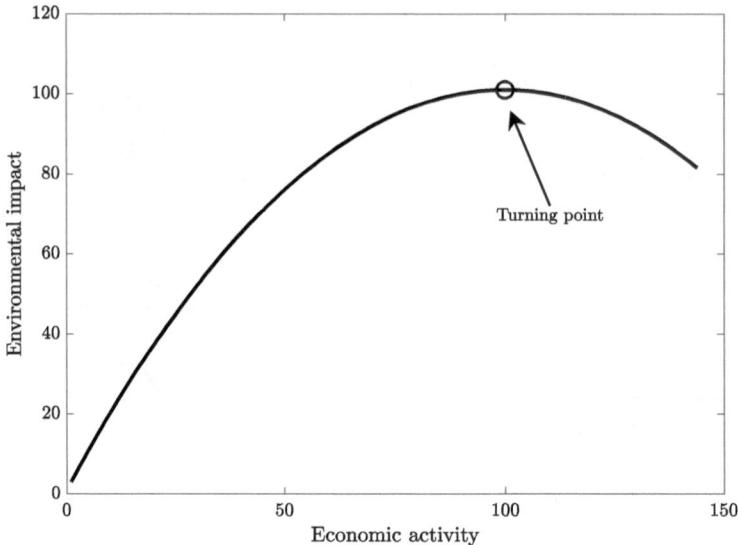

Fig. 16.3 Schematic diagram of an environmental Kuznets curve

16.2 Statistical Analysis

To assess the environmental Kuznets curve hypothesis in its simplest form, the first
question arising is whether a statistically meaningful inverted U-shaped relationship
actually exists between the logarithms of GDP per capita and emissions per capita.
Figure 16.4 shows a scatter plot of the relationship between these two quantities
for the four countries considered in Fig. 16.2. Being generous in the assessment, the
figures display an inverted U-shaped relationship or something similar, although
graphical inspection, of course, is no substitute for a methodologically correct
statistical analysis.

The simplest starting point for statistical analysis of the EKC is a linear
regression model of the form:

$$y_t = c + \delta t + x_t \beta_1 + x_t^2 \beta_2 + u_t, \tag{16.1}$$

with y_t denoting the logarithm of CO_2 emissions per capita in year t and x_t the
logarithm of GDP per capita in year t. In addition, the regression equation includes
a constant and a time trend. The unobserved quantity u_t is the error term. The shape
of the curve, in particular its maximum, the so-called turning point (see Fig. 16.3 for
a schematic representation), is important in applications. If one assumes the above
relationship—with x_t given as the logarithm of GDP per capita in US dollars in 2011
prices in our application—the turning point occurs, when translated to US dollars
GDP per capita (in 2011 prices), at $\exp\left(-\frac{\beta_1}{2\beta_2}\right)$.

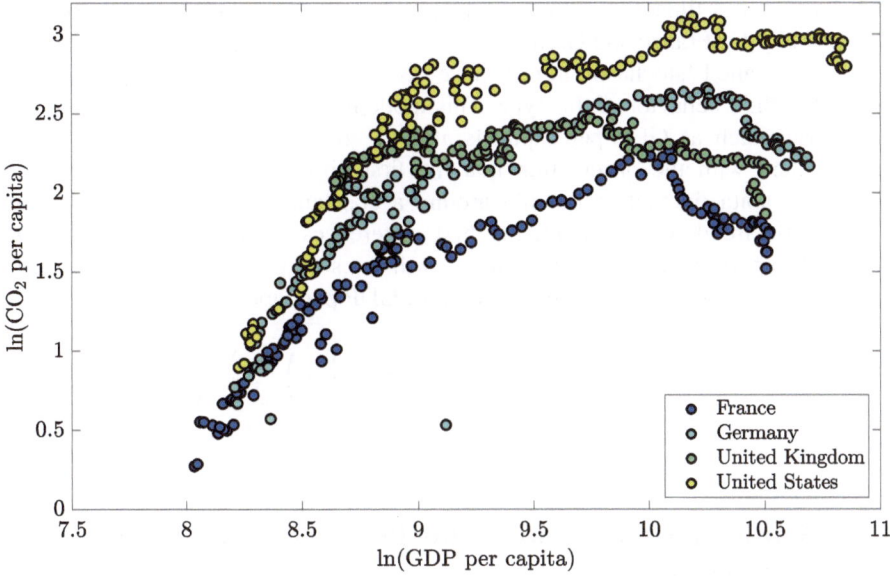

Fig. 16.4 Natural logarithm of GDP per capita in relation to the natural logarithm of CO_2 emissions per capita over the period 1870–2014

Due to the characteristics of the data observed, the linear regression model (16.1) is not easy to analyze, i.e., it is a non-trivial problem to estimate the unknown parameters c, δ, β_1 und β_2 with favorable properties. The notion of favorable properties refers to two aspects: First, consistency, i.e., with the sample becoming larger—and thus more information becoming available—the estimated parameter values move closer to the unknown true values. This is typically a minimum prerequisite for parameter estimation to ensure that the estimated parameter values end up 'close' to the unknown true values. It is, for instance, obvious that only when the estimated parameters are close to the true values, the estimated turning point, too, will be close to the unknown true turning point. Second, it is important to estimate the parameters in a way that allows to test hypotheses, e.g., whether $\beta_2 = -0.5$ or, more generally, whether β_2 is negative, defining an inverted U shape rather than a U shape.

Being a linear regression model, the parameters of Eq. (16.1) would typically be estimated by the ordinary least squares method. Ordinary least squares is the most commonly used parameter estimation method in linear regression models. In the present situation it is not an appropriate choice, however, because the data display properties that complicate statistical analysis and cause ordinary least squares to exhibit unfavorable properties.

Taking another look at Fig. 16.2, it is clear that the GDP-per-capita time series display upward trends while the emissions-per-capita time series possibly feature an inverted U-shaped trend component. However, such trends cannot be described adequately by means of a linear or quadratic time trend as there are large fluctuations and changes over time in the slopes of the trends, with large dips during World War II, especially for France and Germany.

Such so-called 'stochastic trends' have become widespread for describing randomly trending behavior of this type. This modeling approach for macroeconomic time series such as GDP per capita is also of great practical relevance since it is consistent with the observation that the first differences of the logarithm of GDP per capita fluctuate randomly around a constant mean value, i.e., without a stochastic trend. Furthermore, the first differences of the logarithm of GDP per capita approximately describe the growth rates of GDP per capita, i.e., with $x_t = \ln(X_t)$ denoting the natural logarithm of GDP per capita:

$$x_t - x_{t-1} = \ln(X_t) - \ln(X_{t-1}) = \ln\left(\frac{X_t}{X_{t-1}}\right) \simeq \frac{X_t - X_{t-1}}{X_{t-1}}.$$

16.3 Parameter Estimation in the Presence of Nonlinear Cointegration

Modeling meaningful relationships between stochastically trending time series is known in the literature as 'cointegration analysis'. Clive Granger and Robert Engle were awarded the Nobel Memorial Prize in Economic Sciences in 2003 for developing cointegration analysis among other contributions. The results of two statistical tests for testing the hypothesis that the above model correctly describes the relationship between GDP per capita and emissions per capita can be found in the last two columns of Table 16.1. The null hypothesis of the test $P_{\hat{u}}$ is that the error term exhibits a stochastic trend versus the alternative hypothesis that the error term does not exhibit a stochastic trend. The null and alternative hypotheses of the test called CT are the exact opposite, i.e., the error term does not exhibit a stochastic trend under the null hypothesis. Values of test statistics in bold print indicate that the respective null hypothesis is rejected at the 5% level of significance.

According to the results in Table 16.1, strong evidence for an environmental Kuznets curve of a form as given in Eq. (16.1) with an error term that does

Table 16.1 Estimations and test results. Parameter estimation is performed using the FM-CPR estimator of Martin Wagner and Seung Hyun Hong (cf. Sect. 16.5). The last two columns show the results of two tests for the existence of a quadratic EKC. Bold print indicates significance at the 5% level

Country	$\hat{\delta}$	$\hat{\beta}_1$	$\hat{\beta}_2$	Turning point	$P_{\hat{u}}$	CT
France	**−0.004**	**11.11**	**−0.56**	20,544	28.33	0.07
Germany	−0.001	**10.80**	**−0.54**	21,950	**68.75**	**0.11**
United Kingdom	**−0.005**	9.15	**−0.46**	21,043	**90.12**	0.08
United States	0.003	**10.30**	**−0.52**	21,565	12.79	**0.15**

not contain a stochastic trend, can be found for only one of the four countries considered, namely the United Kingdom. For this country, the null hypothesis is rejected by the first test, but not by the second. By contrast, there is strong evidence against the existence of a quadratic EKC for the United States: The null hypothesis is not rejected by the first test, but it is rejected by the second.

The tests lead to contradictory results for France and Germany, with both tests rejecting the respective null hypothesis for Germany and both tests not rejecting the null hypothesis for France. The contradictory results of the two tests are unsatisfactory but, at the same time, an indication that the problem should be studied more closely: The environmental Kuznets curve might not be quadratic. There might be changes in the parameters over time. So-called outliers might bias the behavior of the tests, like the World War II data for Germany, for instance. Additional explanatory variables might be missing, potentially reflecting the three mechanisms mentioned above whose interaction determines the existence and shape of an environmental Kuznets curve. Therefore, all these aspects should be addressed in a comprehensive statistical analysis.

It is well known that in small samples both tests tend to reject a correct null hypothesis too frequently. Therefore, the contradicting test results lead us to tentatively conclude that for both France and Germany—in addition to the United Kingdom—a quadratic environmental Kuznets curve might be present. This decision finds some, albeit only visual, support when considering the estimated error terms, the so-called 'residuals', in Fig. 16.5. There are large outliers for Germany in and around World War II. For France and the United Kingdom, the residuals mostly fluctuate around zero without visible trends. For the United States, the residuals exhibit an inverted U over roughly the first two thirds of the period considered. It

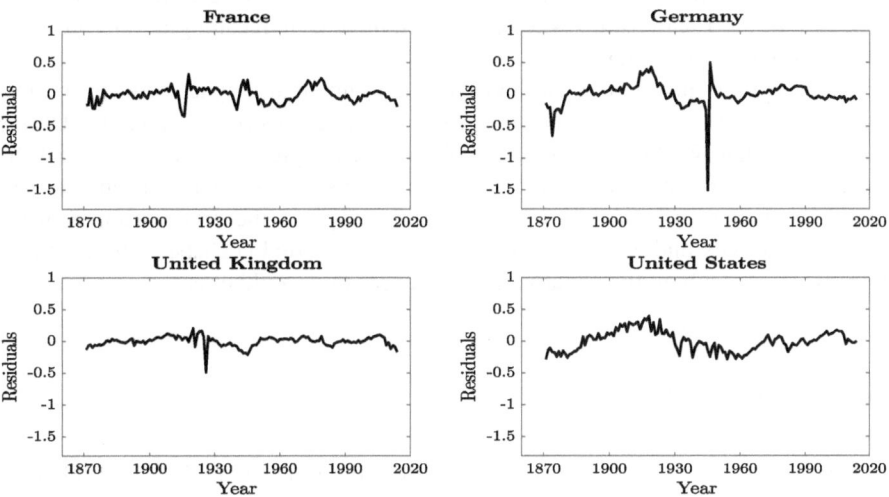

Fig. 16.5 Estimated error terms (residuals) from the estimation of Eq. (16.1)

might, therefore, be useful to estimate an environmental Kuznets curve including additionally higher powers of x_t for the United States.

16.4   Interpretation

Finally, what can be said about the estimation results for France, Germany, and the United Kingdom? First, we observe that the time trend for all three countries has a negative coefficient, while the slope is significantly different from zero for France and the United Kingdom only. The literature assumes that there is an autonomous tendency towards increasing energy and resource efficiency, which is typically captured by a linear time trend. This is the aforementioned technological effect. In Germany, this effect is—in our simple specification—not significant. Moreover, the parameters of GDP per capita and its square are significant throughout and the estimated coefficients to the square of GDP per capita are negative. Consequently, an inverted U-shaped relationship is modeled for all three countries. Finally, the model-based turning points are very close to each other with respect to the GDP-per-capita value but have been reached at different points in time, i.e. in Germany in 1971, in France in 1973, and as late as 1984 in the United Kingdom.

It should also be noted that the results for the United States are very similar to those for the other countries. In a certain sense, it is unavoidable that—even when there is no quadratic environmental Kuznets curve—applying least squares-type methods leads to a 'promising approximation'. To put it somewhat simply, we are basically drawing the one curve—in our case a quadratic curve—through the scatter plot that corresponds to the minimum sum of squared distances of the points in the scatter plot from this curve. A 'best curve' in this sense, therefore, also exists in the case of spurious relationships where, however, there is no meaningful interpretation of the estimated parameters.

Is everything just fine then and the environmental impact will from now on decrease 'automatically', so to speak, in France, Germany, and the United Kingdom? That would be a dangerous fallacy. If people came to believe that the trend inevitably points towards decreasing emissions and one can do as one pleases, this very change in behavior would cause a change in the relationship between economic activity and environmental impact. The inverted U shape could then turn into a U shape with all the negative consequences this would entail.

16.5   Further Reading

A detailed description of the used FM-CPR estimation method of the unknown regression parameters developed by Martin Wagner and Seung Hyun Hong (2016) is contained in "Cointegrating Polynomial Regressions: Fully Modified OLS Estimation and Inference," Econometric Theory 32, 1289–1315.

Part IV
Nature and Technology

Chapter 17
Flood Statistics: Still on the River Bank or Already in the Water?

Svenja Fischer, Roland Fried, and Andreas Schumann

Abstract Floods cause enormous damage. Despite decades of research, they still surprise us in their impact, and a better understanding of their underlying causes and processes would be of huge humanitarian and monetary importance. Statistics provides models and methods for analyzing the risk of floods and, more generally, extreme natural events. In the present chapter, we report on the state of research, ongoing developments, and open problems in the field of research on severe weather conditions.

17.1 Getting a Grip on Floods

In August 2002, large parts of Dresden, the capital of the German federal state Saxony, were flooded (cf. Fig. 17.1). Many buildings and traffic routes, newly built or renovated since the German reunification in 1990, were destroyed, and essential parts of Dresden's world cultural heritage were threatened. The famous *Zwinger*, for example, was flooded for the first time since 1845, as were the *Semper* Opera House and the neighboring depots of the State Art Collections. Damages to the *Semper* Opera House alone amounted to 27 million euros. In total, flood damages in Germany in 2002 amounted to 9.1 billion euros, of which only 1.8 billion had been insured.

These floods in Dresden had not been expected. Since the commissioning of the *Vltava* cascade in the mid-twentieth century, a series of dams in the then Czechoslovak Republic, Dresden had remained largely flood-free. The inhabitants felt safe; the last similarly extreme flood dated back to 1890 and public fears had

S. Fischer · A. Schumann
RU Bochum, Engineering Hydrology and Water Resources Management, Bochum, Germany
e-mail: svenja.fischer@ruhr-uni-bochum.de; andreas.schumann@hydrology.ruhr-uni-bochum.de

R. Fried (✉)
TU Dortmund, Department of Statistics, Dortmund, Germany
e-mail: msnat@statistik.tu-dortmund.de

© The Author(s), under exclusive license to Springer-Verlag GmbH, DE,   143
part of Springer Nature 2024
C. Weihs et al. (eds.), *Statistics Today*, Society, Environment and Statistics,
https://doi.org/10.1007/978-3-662-68907-3_17

Fig. 17.1 Elbe flood in August 2002, which, among other things, flooded large parts of Dresden's city center. The picture shows the *Münzgasse* (© own photo. All Rights Reserved.)

disappeared since then. This changed with the 2002 Dresden flooding and, even worse, once the damage of the 2002 disaster had been repaired, the Elbe burst its banks again in 2013. This new major flood was weaker than the one in 2002, but the water level again exceeded the peak of 1845.

Apparently, floods occur rather erratically. This was also experienced by the citizens of Cologne, whose old town was flooded twice in quick succession in December 1993 and in January 1995. Cologne had experienced similarly large floods in January 1920 and in January 1926.

The Rhine floods of 1993 were the beginning of a series of extreme floodings in Germany. After the following Rhine flood in 1995, the Oder was affected in 1997, the Elbe and the Danube in 2002, large parts of Bavaria in 2005, and, again, in June 2013 not only the Elbe and Danube but this time also the Saale. On June 13, 2013, the editor-in-chief of Bavarian television, Sigmund Gottlieb, asked aghast on the German news how it could be possible that the entirety of the worldwide engineering expertise is able to solve almost any problem but not this one; humanity has targeted Mars but is apparently not able to get floodings under control.

Is this really true? How can we get floods under control? And how can statistics help the engineers?

17.2   What Is a Flood?

The Water Resources Act defines 'floods' as the temporary inundation of land that is normally not covered with water. Since those responsible for flood protection naturally want to prevent floodings, such an event-oriented definition is not sufficient.

In hydraulic engineering, the notion 'flood' refers to the temporary exceeding of a threshold of the discharge at a certain watercourse cross-section. The discharge is the volume of water that flows through a river section in a certain unit of time, usually one second. This discharge fluctuates temporally. In Germany, it is particularly high in spring and usually decreases towards autumn. These differences strongly depend on the size of the catchment area that drains into the respective river up to the gauging station under consideration. The mean low flow discharge (MNQ) and the mean peak flow discharge (MHQ) (cf. Fig. 17.2) differ by a factor of 7 for the Rhine at the Cologne gauge (about 144,000 km^2 catchment area). For the Elbe in Dresden with a catchment area of 53,000 km^2, this difference is substantially larger with a factor of 18, and for Bad Düben at the Mulde (catchment area 6200 km^2) the factor is as high as 119. The annual flood discharges also vary. In Cologne, the largest measured flood in 1926 was 1.8 times higher than the mean value of

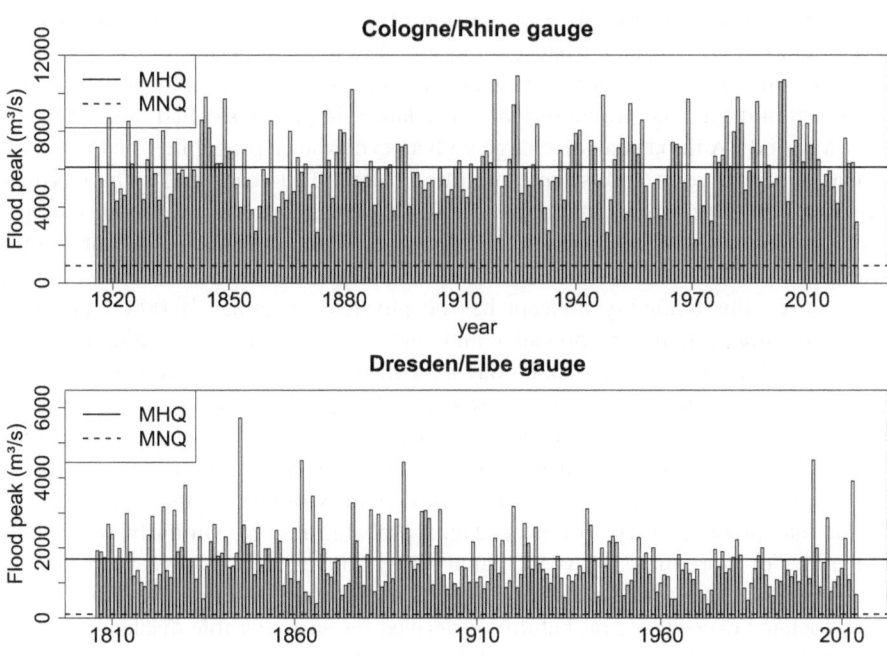

Fig. 17.2 Time series of the annual maximum discharges for the Cologne/Rhine and Dresden/Elbe gauges, as well as their mean low flow discharge (MNQ) and mean peak flow discharge (MHQ)

the annual maximum discharges; in Dresden it was 2.8 times this value and in Bad Düben 4.7 times, both in 2002.

17.3   Flood Risk and Probabilities

On what basis can flood protection be planned and the flood risk assessed?

In order to use the fertile, but flood-prone floodplains, these areas have been protected by dikes over the centuries (for example, on the Lower Rhine for about 800 years). Floodings thus became less frequent, people dared to erect buildings, the monetary values of dike-protected areas increased, and the protective measures were expanded. However, people had to recognize that ongoing improvement in flood protection was very costly and that complete protection was not possible, for economic reasons. Therefore, the exceedance probability of the flood peak discharges has become the basis for the assessment of flood risk and flood protection.

To assess the flood risk at a gauging station, one can analyze the highest flood discharges from earlier years. The highest observed value of a year in an observation series of, for example, 100 years is reached only once and is not exceeded 99 times. The relative proportion of the highest observed value is thus at 1%. Assuming constant basic conditions, this relative proportion is a simple estimate of the probability of a flood to occur of at least this size within any year. The reciprocal of the annual probability of exceedance (in years) is called 'annuality' or 'return period'. A flood peak discharge with an exceedance probability of 1% has an annuality of 100 years and is therefore often called '100-year flood'. The German DIN norm 19700 specifies that dams should be designed for maximum discharges with a probability of exceedance of 0.01%, corresponding to an annuality of 10,000 years.

However, this annuality concept has no physical meaning. 10,000 years ago, different climatic conditions prevailed and our region was still in transition between the Weichselian glacial period and today's warm stage. In the same vein, the climate in Germany would not be exactly the same in 10,000 years as it is today. Annuality is simply a notion of probability that has been in use for the last 100 years. It ultimately does not reveal anything about the intervals at which major flood events actually follow one another, as the examples for Dresden and Cologne exemplify.

Put into practice, very low exceedance probabilities, or, similarly, very high annualities of more than 100 years, cannot be estimated from 100 years of data or less. Therefore, statistics works with models of distribution functions, from which the associated exceedance probability is derived for each possible discharge value–provided one has chosen the right model. Often, theoretically justified assumptions are made as to what the distribution function might look like. This reduces the problem of fitting a distribution function to the observed annual maxima to the need to estimate a few unknown numerical values from the data only. These numerical values are called model parameters and completely describe the distribution, which was established under the particular theoretical assumptions.

The annual maximum is the maximum of many individual values. For this reason, the Fisher-Tippett-Gnedenko theorem is often employed when modeling the annual maximum values. According to this theorem, the maximum of many independent, identically distributed individual values follows approximately a so-called Generalized Extreme Value distribution under certain mathematical conditions (cf. Chap. 7). To fit such a Generalized Extreme Value distribution, one only needs to estimate three distribution-specific parameters from the data. These are called location, scale, and shape parameters and determine the mean level, the variability, and the speed of decay of the exceedance probabilities for increasingly large values, respectively. Since estimation of the shape parameter from short observation series is particularly difficult, this parameter is often set somewhat arbitrarily to the value 0 for simplification, which yields the Gumbel distribution as a special case of the Generalized Extreme Value distribution. As an alternative to the Generalized Extreme Value distribution, similar models such as the Pearson-III distribution are commonly used.

There are various statistical methods for estimating the model parameters, of which in hydrology so-called L-moments, or alternatively the method of product moments or maximum likelihood estimators, are most popular. For the design of a dam, for example, the 99.99% quantile of the estimated distribution model is used, i.e., the value that is only exceeded with a probability of 0.01% if the model is valid. While the estimated quantiles for small and medium values are generally quite similar, large differences between the distribution models and estimation methods occur in the right tail, i.e., for extreme floods. While for an annuality of 100 years the estimated quantiles may differ by 30–40%, the differences can become much larger for higher annualities (see Fig. 17.3). Often one tries to select an appropriate calculation method by comparing the calculated values with the observed frequencies.

Up to now, when investigating annual maximum flows, we have limited ourselves to a single maximum value per year, i.e., other floods in the same year have not been taken into account. However, some of these floods might have been larger than the annual maximum flows in other years. It is therefore possible that the maximum annual flows do not cover the full set of largest floods.

Another approach is to use all values above a predefined threshold for the statistical analysis instead of the annual maximum flows. This approach is known as peaks-over-threshold (POT). For the POT approach, however, the choice of the data basis is unclear. Values on consecutive days should not be used as they are highly dependent. Monthly peak flows can also be dependent, since the maximum peak of the next month may occur at its beginning and can still belong to the flood event of the preceding month. Such dependencies must be excluded in each individual case since each flood should only be counted once. The choice of the threshold value is also problematic. It can be chosen statistically, e.g. via the method of the mean-residual-life plot, or via a fixed, average number of values per year. In some cases,

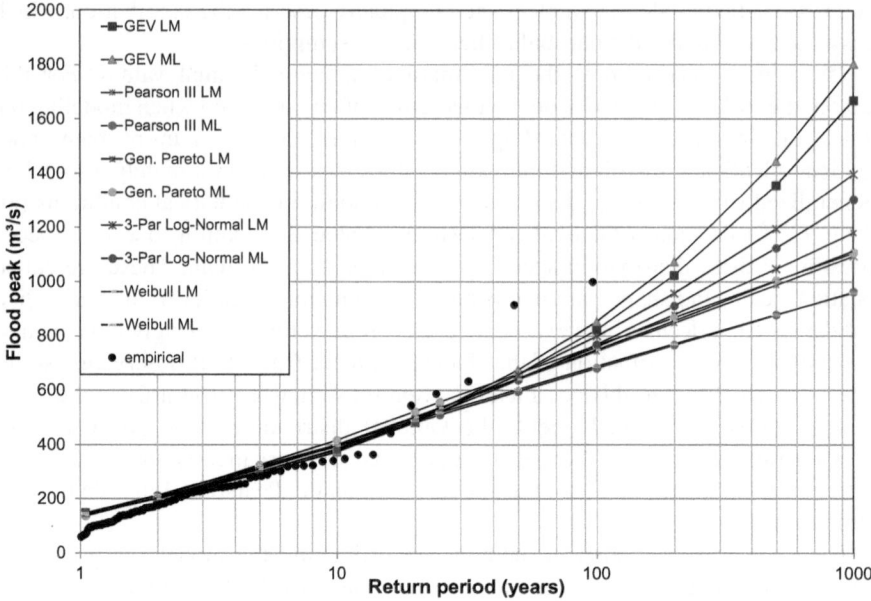

Fig. 17.3 Empirical non-exceedance probabilities (points) as well as distributions fitted with different parameter estimation methods

the threshold for floods is determined hydrologically as two or three times the mean discharge.

Finally, in order to infer the annuality from the distribution of flood peaks above the threshold, the exceedances per year are modeled by means of a so-called Poisson model, and the POT and the Poisson model are combined to estimate future annual maximum flows.

17.4   Robust Estimation

New annual maxima or threshold exceedances change the estimates of the quantile values that enter the design of, e.g., a flood protection dike. A single major flood may majorly increase the estimates of the quantiles of the distribution, given the often short series of observations of 50 or fewer years. As a result, dikes—even if they resisted the flood—may have to be raised as they may be regarded as too low. If subsequently no further major floods occur for a long period of time, the estimated exceedance probability decreases, only to increase again after the next major flood.

Such fluctuations can be reduced by using new, robust estimation procedures for the parameters of the distribution function. Robustness in the statistical sense means that a single extraordinarily large or small value has only a restricted influence on the estimation results such as the fitted distribution function or the resulting annuality. Therefore, new observed values can improve our estimates, but one large flood event cannot drastically change them.

The best-known example of a robust estimator for the mean location of data is the median. While the arithmetic mean is strongly influenced by individual extreme values, the median corresponds to the value that is in the middle of the observation series after re-ordering by size. The maximum value does not directly influence the median as it is just one of the large values of a distribution and its exact value is irrelevant for the computation of the median.

As an alternative, a trimmed mean can be used as a compromise between median and arithmetic mean, where the arithmetic mean of, e.g., the middle half or the middle 90% of the data is calculated. Trimmed means are often used in sports to summarize judges' scores because the highest and lowest scores are not necessarily meaningful for determining the mean score.

Flood statistics, of course, require more sophisticated methods. The simplest example is the trimmed L-moments (TL-moments), which, similar to the trimmed mean, reduce the influence of extreme individual values via a trimming factor. Despite this trimming, extreme events are not systematically underestimated because the existence of the trimmed most extreme values is taken into account, but not their exact value. Simulations of artificial flood time series confirm that the short sample size and thus the great influence of very rare floods are the cause of the high fluctuations in the estimation results of non-robust methods. As the observation series get longer, the estimates based on, e.g., L-moments drop down to the level of the robust estimates, as the example in Fig. 17.4 shows.

17.5   Flood Types and Changes Over Time

Flood probabilities can vary over time. This does not refer to the changes of estimated values due to new observations discussed in the previous section, but rather to the consequences of climate fluctuations or human intervention. Among other causes, river engineering measures can accelerate flood discharges in such a way that waves from tributaries unfavorably coincide with the flood wave of the main river. The expansion of the Upper Rhine, for example, causes its flood peaks to overlap unfavorably with those of the Neckar. As a result, extreme flood peaks have become more frequent, so that the exceedance probabilities calculated previously are too small. The effects of such changes can be taken into account in the calculations by correcting the observation series so that it corresponds to more recent conditions.

Climatic conditions have also influenced the occurrence of floods. Twelve major flood events that have occurred simultaneously on the Danube, Rhine, Elbe, and Weser since 1920 always occurred in the months of January, February, or March. In each of these cases, the cause was snowmelt, often combined with ice jams. Note that the frequency of snow cover in Germany has varied greatly between decades.

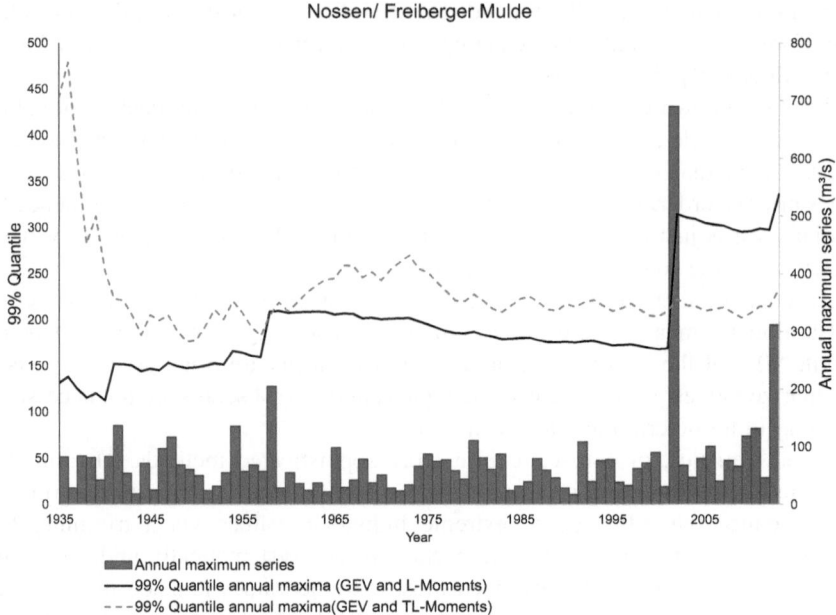

Fig. 17.4 Annual maximum flows of the Nossen/Freiberger Mulde gauge and estimated 99%-quantiles using L-moments and robust TL-moments

In addition to snowmelt, floods can also be caused by rain. Heavy rainfalls lasting for several hours with high rain intensity can lead to a short but heavy flood wave. An equally high peak discharge can also occur as a result of several days of continuous rain, but with a significantly larger volume of the flood wave.

Snowmelt, on the other hand, leads to very long-lasting floods with a very large volume. In Germany, flood events occur most frequently due to continuous rainfall, while floods caused by heavy rainfall occur less frequently and mostly in the summer months, often in conjunction with thunderstorms. These seasonal differences put the standard model of a Generalized Extreme Value distribution for the annual maximum values into question, as it is theoretically only justified for maxima of values with identical distributions, without such systematic seasonal differences.

In order to take these differences into account, we can consider heavy rain, continuous rain, and snow melt events separately and fit separate distributions to the extreme observations for each of these flood types within the framework of the POT approach (Fig. 17.5). For a statement on the annuality of a flood of any type, we combined these distributions via a so-called mixture model, which is an additive combination of these distribution functions with weightings according to the relative frequency of the flood types.

The flood types also differ in their seasonal occurrence. The example of snowmelt floods makes it clear that not every flood type occurs with the same frequency in

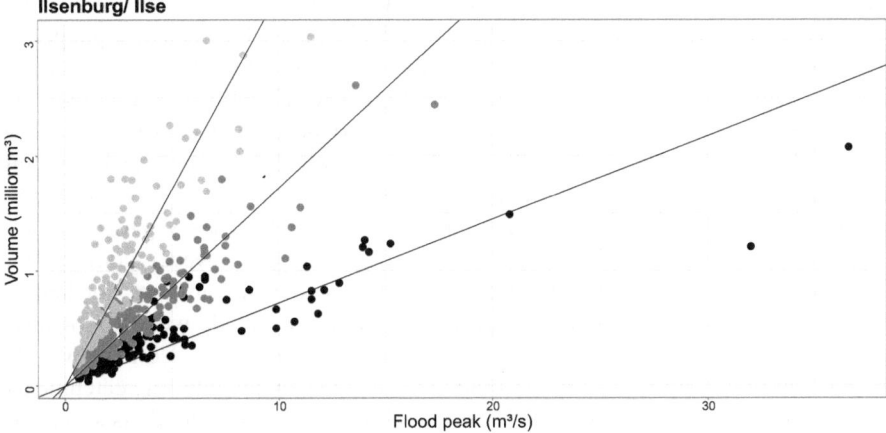

Fig. 17.5 Typing of floods into heavy rainfall events (black), continuous rainfall events (gray) and snow (melting) events (light-gray) on the basis of their peak-to-volume ratio for the Ilsenburg/Ilse gauge

every season. Moreover, the characteristics of a flood can also vary seasonally due to different climatic conditions. If precipitation falls on an already wet soil, as it is often the case in spring, this leads to faster runoff because less water can be absorbed by the soil. The flood then tends to be larger than during a dry period when the soil has more absorbing capacity. In addition to the condition of the soil, there are many other seasonal factors of influence, so that a seasonally differentiated analysis is also indicated. Differentiation with respect to both flood type and seasonality is potentially more informative than a simplistic joint consideration of all observations, neglecting their heterogeneous formation conditions.

17.6   Regionalization

For most gauging stations, only comparatively few, and in some river regions even no, measured values are available for estimating the model parameters. This problem can be overcome through regionalization, i.e., the transfer of information gained from data measured at several other catchment areas that are hydrologically comparable to the station we are interested in. Joint estimation from catchment areas with similar statistical properties allows for more precise statements about exceedance probabilities at gauges for which little or no data are available.

Of course, floods at different gauges do not behave exactly the same. The much-used 'index flood' assumption states that the flood distributions of a group of similar catchment areas differ only by a scaling factor. Application of this idea requires finding homogeneous groups of similar catchment areas in a first step, and then a joint estimation of the common model parameters, which are assumed to be identical within each group. Differences between the gauges in the same group are accounted for by estimating the individual scaling factors using specific catchment characteristics such as size, elevation, or land use. For the grouping, we use expert knowledge of experienced hydrologists as well as statistical clustering methods based on the location, soil characteristics, surrounding vegetation, and other characteristics of the individual gauges. In the joint estimation, we must take into account that 'simultaneous' floods at close-by gauges should not be treated independently as they depend on each other. However, conventional correlation-based approaches describe average dependencies and are not helpful for the joint analysis of dependent extreme values. Instead, we use so-called extreme value copulas to model dependencies of extremes. This also improves the assessment of the risk that is due to simultaneous floods in neighboring catchments. Overall, however, the assessment of the magnitude of extreme floods is subject to considerable uncertainty and will continue to be a major challenge in the future due to the many factors which influence the emergence of large floods, as discussed in this chapter.

17.7   Further Reading

The first systematic paper on flood statistics is "The return period of flood flows" by Emil Julius Gumbel, published in the Annals of Mathematical Statistics in 1942. The statements on robust statistics in hydrology in Sect. 17.3 are mainly based on Fischer, S., and Schumann, A., "Robust flood statistics – comparison of peak over threshold approaches based on monthly maxima and TL-moments", Hydrological Sciences Journal 61, 2016, 457–470. The type- and seasonally-differentiated flood statistics from Sect. 17.5 can be found in Fischer, S., Schumann, A.H., and Schulte, M., "Characterisation of seasonal flood types according to timescales in mixed probability distributions", Journal of Hydrology 539, 2016, 38–56, and in Fischer, S., "A seasonal mixed-POT model to estimate high flood quantiles from different event types and seasons", Journal of Applied Statistics, DOI: 10.1080/02664763.2018.1441385. More on regionalization (Sect. 17.6) can be found, e.g., in Lilienthal, J., Fried, R., and Schumann A.H., "Homogeneity testing for skewed and cross-correlated data in regional flood frequency analysis", Journal of Hydrology 556, 2018, 557–571.

Chapter 18
How Statistics Helps to Reduce Rejects

Claus Weihs and Nadja Bauer

Abstract No company likes to produce defective goods (rejects). The Six Sigma method is a recognized statistical approach that helps to reduce reject rates. By means of an example from a deep drilling process, the present chapter demonstrates how statistical methods and modern process management fruitfully interact in such an approach.

18.1 Defects in Deep Drilling

In this chapter, we demonstrate how the statistical approach called Six Sigma (6σ) method can improve an industrial process in interaction with modern process management. We illustrate the effect of this approach by means of the metal-cutting manufacturing process called BTA (Boring a Trepanning Association) deep drilling. By means of this process, holes are drilled with bore diameters of about 18–2000 mm to produce holes with a length of three to 100 times their diameter. Due to this large ratio, this process is called deep hole drilling. The metal chips produced in the BTA process are disposed of through the entrance of the hole using oil as a lubricant.

In our example, we worked with a company which produces workpieces for turbines using the BTA process and which is struggling with a high rate of manufacturing defects and a decreasing number of orders due to customer dissatisfaction. In the drilling process, a problem typical for deep drilling occurs, the so-called 'rattling', which is caused by self-excited torsional vibrations. This rattling creates radial marks, also called chatter marks, at the bottom of the workpiece, which leads

C. Weihs (✉)
TU Dortmund, Department of Statistics, Dortmund, Germany
e-mail: claus.weihs@tu-dortmund.de

N. Bauer
FH Dortmund, Faculty of Computer Science, Dortmund, Germany
e-mail: nadja.bauer@fh-dortmund.de

© The Author(s), under exclusive license to Springer-Verlag GmbH, DE, part of Springer Nature 2024
C. Weihs et al. (eds.), *Statistics Today*, Society, Environment and Statistics, https://doi.org/10.1007/978-3-662-68907-3_18

to a great loss of quality and to increased tool wear. In the following, we show how the BTA process can be improved by means of the Six Sigma method, which aims at the reduction of manufacturing defects to a maximum of 3.4 defects per one million pieces.

18.2 Quality Improvement: Six Sigma

The quality of a technical process is typically measured by means of certain quality characteristics, in our case the mean roughness (depth) of the chatter marks. Typically, tolerance limits are specified for such quality characteristics. If the value of such quality characteristics lies outside the specified range, the product is considered defective. The Six Sigma method aims to ensure that manufacturing defects do not extent a maximum of 3.4 defects per one million pieces. Assuming a normal distribution of quality characteristics, standard deviations characterize the extent of the distribution. To achieve the aim of the Six Sigma method, deviations from the mean which are lower than 6σ (sigma), i.e. 6 times the standard deviation of the quality characteristics, need to lie within the specified tolerance region (see Fig. 18.1).

To achieve this ambitious goal, the Six Sigma method uses the so-called 'DMAIC approach', a data-driven cycle of methods for analyzing, optimizing, and stabilizing economic processes. This approach follows five steps, i.e. Define, Measure, Analyze, Improve, and Control (cf. Fig. 18.2). In the following, we apply this cycle to improve the deep drilling process in our example.

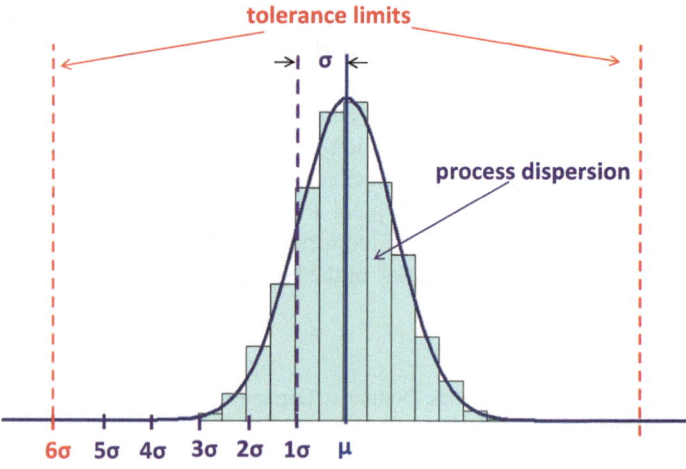

Fig. 18.1 Illustration of the 6σ approach by means of a normal distribution: deviations of more than 6σ from the mean have a probability of 0.00034%

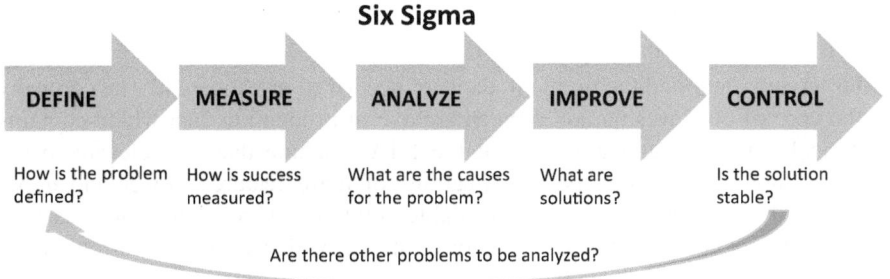

Fig. 18.2 The DMAIC cycle

18.2.1 Define

As a first step, we need to define the problem, in our case the process of BTA bore hole production. This process typically follows five steps: receipt of an order, ordering of drilling heads (if necessary) and unmachined workpieces, execution of the drilling by means of the BTA deep hole drilling machine, checking the quality of the manufactured workpieces, and delivery of the products to the client.

In step four (checking the quality of the manufactured workpieces), manufactured workpieces pass through a quality control machine, which sorts out the defective workpieces. In our example, every 20th manufactured workpiece has minor or significant defects, even though the BTA process is always the same. Still, clients have their own quality controls and occasionally complain about (parts of) the delivery.

The producers have identified three possible causes of the defects:

(a) **suppliers**: the quality of the drill heads varies between the different suppliers;
(b) **machine settings**: the quality of the workpieces depends on the machine settings;
(c) **quality control**: the quality control machine identifies parts as defective that are not, and vice versa.

To reduce rejects, a first step would be prioritization, i.e. deciding which possible cause to focus on. Various management methods exist for making this decision. We do not present these management methods in detail, since the exact processes are not important for the outcome of prioritization. For our example, we could identify the inadequate machine settings (reason b) as the main source of the problem. Therefore, improvement of the machine settings was our first aim. To resolve this issue, we created a so-called 'project profile', which provides an overview of the project and the problem identified, in our case the inaccurate machine settings, as well as the aim of the project, i.e. to specify the minimum quality of the workpieces in order to ultimately decrease the amount of rejects.

18.2.2 Measure

After having identified the problem, data are needed to investigate it. The company at the core of our investigation runs an archive of production data which contains historical data on four parameters of the BTA machine that are relevant to the project. Table 18.1 shows the specifications of these features (lower and upper limits). Table 18.2 presents the 20 available combinations of their values as well as the corresponding proportions of bore holes in which chatter occurred (reject rate).

To illustrate the effect of changing the values in the different parameter combinations of the machine settings and the suppliers (cf. reasons a and b), we split the test results into acceptable (if the reject rate is less than 0.01) and unacceptable (otherwise) outcomes. Small reject rates are therefore acceptable. In this way, 7 of the 20 test conditions (parameter combinations) are identified as acceptable

Table 18.1 Parameters of the BTA deep drilling machine, LL = lower limit, UL = upper limit

Abbreviation	Designation	LL	UL
vc	Velocity of cutting	60 m/min	120 m/min
f	Feed rate	0.05 mm/sec	0.25 mm/sec
oil	Oil pressure	150 bar	450 bar
b	Drill head supplier	Values 1, 2, 3 for suppliers A, B, C	

Table 18.2 Extract from the production data archive (acceptable reject rates in bold)

vc	f	oil	b	Reject rate
120	0.25	450	1	0.012
120	0.05	450	3	**0.003**
60	0.005	450	2	**0.006**
90	0.15	173	1	0.089
64.5	0.15	300	2	0.016
90	0.235	300	3	**0.009**
120	0.05	150	1	0.136
90	0.065	300	1	0.032
120	0.25	150	2	0.100
120	0.25	450	2	**0.006**
60	0.25	150	2	0.090
120	0.05	150	3	0.039
60	0.25	450	3	**0.003**
60	0.25	150	3	0.045
90	0.065	300	2	0.017
64.5	0.15	300	3	**0.009**
60	0.25	450	1	0.012
120	0.05	150	2	0.084
90	0.15	300	2	0.016
90	0.15	300	3	**0.009**

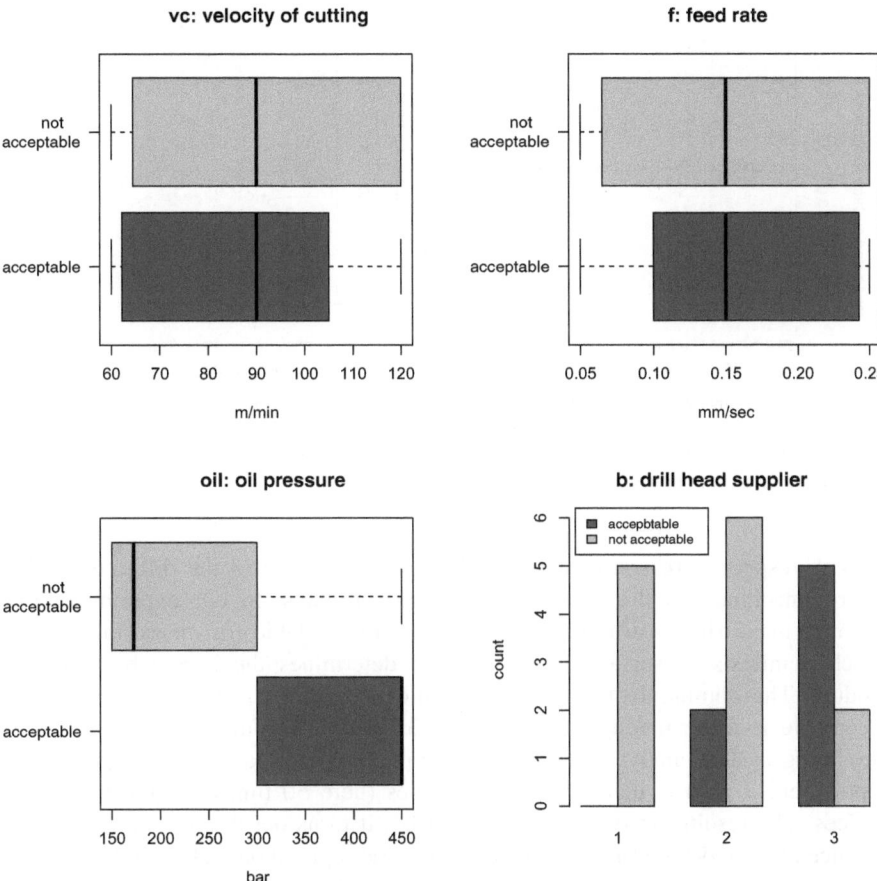

Fig. 18.3 Dependence between the influencing characteristics and acceptance

(acceptable reject rates in bold in Table 18.2). Figure 18.3 illustrates the distribution of acceptable and unacceptable values of the parameters by means of boxplots and bar charts.

It can be seen that for the different values of vc and f there seem to be hardly any differences with regard to the acceptability of the products. On the other hand, all production results are unacceptable at an oil pressure level lower than 300 bar. Furthermore, the third supplier of the drill heads seems to deliver particularly good products. However, these findings may be distorted by the fact that the influencing characteristics are correlated. For a reliable determination of the effect that the different parameters have on product quality, a statistical experimental design (cf. Sect. 20.2 for an introduction) is necessary as illustrated in the next subsection (Analyze).

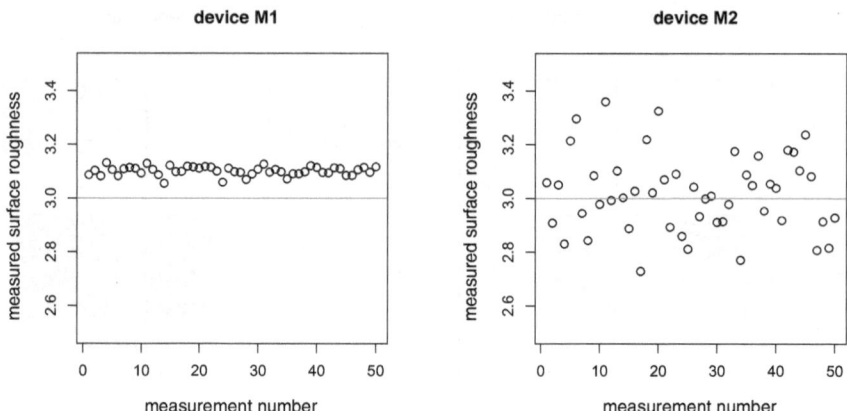

Fig. 18.4 Results of the roughness measurements (in μm) for the devices M1 and M2; green line = envisaged mean roughness depth

With respect to reason c, we considered the influence of the different quality control machines on the acceptability of the products. In our experiment, two different measuring instruments (M1, M2) were available for measuring chatter, which manifests as surface roughness and determines the acceptability of the product. The commonly used upper specification limit for a product to still be acceptable is a mean roughness depth of 3.2 μm (micrometers). In a so-called measuring system analysis, a sample workpiece with a known mean roughness depth (here 3 μm) is measured several times (here 50 times) by the measuring devices. The results are displayed in Fig. 18.4. It turns out that the measurements produced by the device M1 are systematically too high (about 0.1 μm). In contrast, M2 is exact in that it measures a mean value of 3 μm. However, the device M2 measures with a higher variation than M1; this variation is considered unacceptable. After a modification suggested by the manufacturer, the systematic inaccuracy produced by M1 could be resolved and the company decided to use M1.

18.2.3 Analyze

To show how the surface roughness depends on the four parameters introduced in the previous subsection (Measure), our next aim was to optimize the process by means of modeling the surface roughness as a function of the characteristics vc, f, oil, and b with the help of statistical experimental design. For the characteristic b, it has to be noted that, in the meantime, supplier B had moved the production location, so that, for economic reasons, only the two suppliers A and C could be considered.

To test the significance of the independent influence of each of the four parameters on the target variable (i.e. surface roughness), so-called 'screening'

Table 18.3 Plackett-Burman experimental design for the characteristics vc, f, oil (− = LL, + = UL, see Table 18.1) and b (− = supplier A, + = supplier C), and the corresponding mean value of the target parameter roughness (in μm)

Experiment	vc	f	oil	b	roughness
1	−	−	−	−	39.67
2	+	−	+	−	8.52
3	−	+	+	+	22.04
4	+	+	+	+	6.71
5	−	−	−	+	38.43
6	−	+	+	−	23.28
7	−	−	+	+	21.63
8	+	−	−	+	24.08
9	+	−	+	−	8.52
10	+	+	−	+	23.51
11	−	+	−	−	40.08
12	+	+	−	−	24.75

designs of experiments are particularly suitable. In such designs, only two levels per characteristic are used. These should take on values as extreme as possible, i.e. they should be equal to LL and UL if realizable (cf. Table 18.1). They are coded by −1 and +1. For each experiment, an experimental design prescribes the values of each characteristic. These values are listed as either − or +.

The best-known screening designs are the 'Plackett-Burman designs'. Table 18.3 shows such a design with 12 experiments as well as the measured mean values of the surface roughness (mean values over 10 productions). Each row of the design corresponds to an experimental setting and each column to an influencing characteristic. The special feature of this design is that the characteristics (columns) are uncorrelated. Only in such designs, the effects of the characteristics can be determined independently of each other.

The data presented in Table 18.3 were subsequently analyzed by means of linear regression. The following conclusions can be drawn from the results:

- Characteristic vc has a strong influence on surface roughness in that the higher the velocity of cutting the lower the roughness,
- characteristic oil has an influence on surface roughness in that the higher the oil pressure the lower the roughness,
- factor b has an influence on surface roughness in that the better workpieces are produced by drill heads from supplier C, and
- characteristic f seems to have no influence on the target value.

For the company this means that they should buy the drill heads from supplier C. For production, we then have to determine the optimal values of the other three parameters. Higher values of the two characteristics cutting speed (vc) and oil pressure (oil) seem to enhance the quality of the workpiece. Since the characteristic feed rate (f) does not seem to have any influence, it is set to the mean value (f = 0.15 mm/sec) within the permissible range. In contrast to the results presented in the previous subsection (Measure), the parameter vc proves to be important. However, the target value was defined differently in the two contexts: reject rate < 0.01 in the previous subsection and surface roughness in the current case.

18.2.4 *Improve*

To optimize the settings of the two influential characteristics cutting speed (vc) and oil pressure (oil), we use a so-called 'inscribed central composite' experimental design. Here, more than two values per characteristic are used in order to determine those values of the characteristics which reduce surface roughness to its minimum. A model evaluation (not to be reported here) has revealed that the following model found by linear regression fits the data very well:

$$\text{roughness} = 114.732 - 0.23656 \cdot \text{oil} - 1.143967 \cdot \text{vc} + 0.000141 \cdot \text{vc} \cdot \text{oil}$$
$$+0.000276 \cdot \text{oil}^2 + 0.004638 \cdot \text{vc}^2.$$

For the optimal values of vc and oil, the model function on the right-hand side of the equation is evaluated. This suggests that for reaching the minimal and thus optimal value of the target variable surface roughness, vc has to be set to 119 m/min and oil to 407 bar, i.e. almost to the upper edge of the permissible range (cf. Table 18.1). With these settings, the surface roughness is 0.54 μm, i.e. almost zero, which would be the ultimately ideal result.

18.2.5 *Control*

In the final step, we use these optimized settings and check how the process behaves over time with 500 new observations with the help of a so-called 'control chart' (cf. Fig. 18.5). The lower and upper control limits in Fig. 18.5 determine the range in which the target value lies with 95% probability if all uncontrolled parameters remain stable. All values outside the control limits are considered to be exceptional and the reasons for this should be discussed. We can see that, in our example (cf. Fig. 18.5), only one out of the 500 values lies outside the control limits, so the process seems to have been out of control for a short period of time. However, since this is only a single value, we can safely ignore this 'outlier'.

Last but not least, a 'process capability analysis' relates the range of variation of the target variable measurements, in our case of the 500 new observations, to the specification limits. From this analysis, we get a probability of 0.001056% that the upper specification limit of 3.2 μm will be exceeded. This means that this happens in about 11 per 1 million cases, which brings us very close to the 6σ requirement of 3.4 per 1 million.

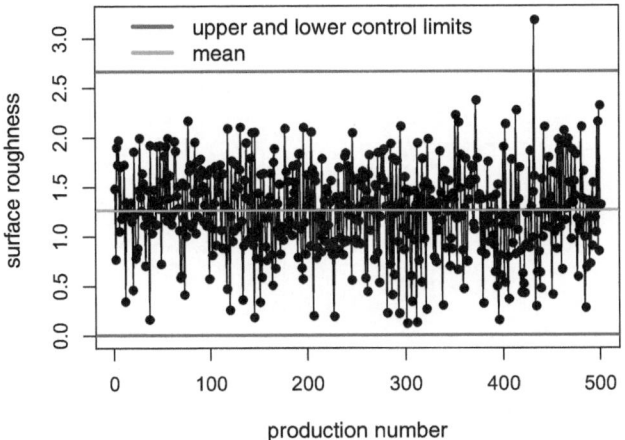

Fig. 18.5 Control chart for the target variable roughness (in μm) with optimal feature settings

18.3   Further Reading

The problem definition and the implemented procedure are based on a research project conducted by the first author at the Faculty of Mechanical Engineering of TU Dortmund University. Further information on the BTA deep drilling process can be found in Webber (2007): "Untersuchungen zur bohrtiefeabhängigen Prozessdynamik beim BTA-Tiefbohren," Volume 39 of the ISF publication series, TU Dortmund. For more information on experimental designs, control charts, and process capability in Sects. 18.2.3, 18.2.4, and 18.2.5 see Sect. 20.2 and Weihs and Jessenberger (1999): "Statistische Methoden zu Qualitätssicherung und Optimierung" (Statistical Methods for Quality Assurance and Optimization), Wiley-VCH, pp. 239, 250, 282, and 350, respectively.

Chapter 19
Statistics and Reliability of Technical Products

Christine H. Müller

Abstract The train disaster of Eschede (Germany) in June 1998 with more than 100 deaths has shown which consequences can occur if a technical product fails. Only one wheel broke due to material fatigue. As a result, many trains were taken out of service for a long time, resulting in traffic chaos. Was that necessary at all? And could the accident have been avoided through better reliability analysis? Statistics have a lot to say about this.

19.1 Reliability and Randomness

For a long time, it was thought that statistics were not needed in technology, since everything is given by the deterministic laws of mechanics. However, randomness plays a bigger role in technology than we might think. For example, durability and wear and tear depend on accidental use, and random climatic factors influence corrosion. Even under constant conditions, technical products age at different rates. For example, material fatigue is often due to micro-cracks that occur and grow by chance.

Crack growth is basically a random process, even if attempts are made again and again to describe it deterministically. In particular, random atomic processes influence the formation and growth of micro-cracks. In addition, there are random micro-structures of the material in which the cracks grow. Figure 19.1 shows the surface of a steel sample photographed through a microscope before exposure and after 5000 and 18,000 cyclic load changes. The resulting micro-cracks, visible through darker values in the image, are distributed totally randomly on the surface and also grow at different speeds. The darker areas that can be seen before exposure are, however, caused by dirt and scratches on the material.

C. H. Müller (✉)
TU Dortmund, Department of Statistics, Dortmund, Germany
e-mail: cmueller@statistik.tu-dortmund.de

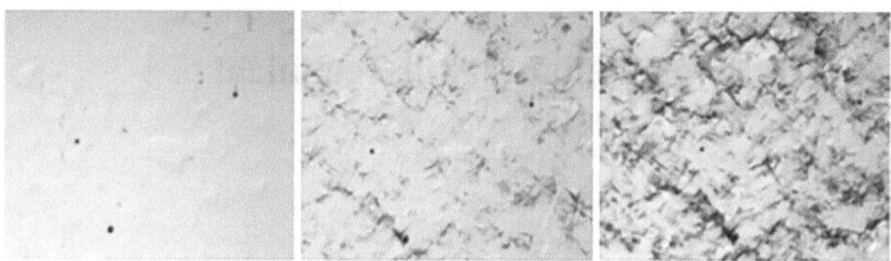

Fig. 19.1 Formation of micro-cracks on a steel sample. Left: surface before exposure. Middle: surface after 5000 load changes. Right: surface after 18,000 load changes (Own graphic with R image: https://www.rdocumentation.org/packages/graphics/versions/3.6.0/topics/image.)

It is therefore clear that the reliability of technical products is strongly influenced by randomness. Only statistical methods can therefore adequately record the durability and service life of technical products and make predictions for their future behavior.

19.2   Simple Service Lifetime Analysis

The simplest model of service lifetime analysis is given by the exponential distribution. According to this, the probability that the lifetime T of a product is greater than a value $t > 0$ is the following:

$$R(t) = P(T > t) = e^{-\lambda t}.$$

The function R is also called the reliability function. Here $e \approx 2.71828$ is the so-called Euler's number and only $\lambda > 0$ is an unknown parameter that indicates the expected (average) lifetime. This expected lifetime $E(T)$ satisfies

$$E(T) = \frac{1}{\lambda},$$

i.e. the larger λ, the shorter the expected lifetime. This is often estimated by the arithmetic mean of the observed lifetimes of the product. In an experiment with steel girders, in which the time is measured until a crack occurs, the measured mean lifetime could be 2 years = 730 days. Here the estimated parameter would be $\hat{\lambda} = 1/730$. This means that the probability that a steel girder will remain free of cracks for at least one year is

$$P(T > 365) = e^{-\hat{\lambda} \cdot 365} = e^{-\frac{365}{730}} \approx 61\%.$$

Sometimes, however, not all failure times are available because the experiment cannot run arbitrarily long. Then you just know that the products have not failed by the end of the experiment. Such cases could be left out in the statistical analysis. It is better, however, to include them as so-called 'censored observations' in the statistical analysis, since they also contain information about the lifetime of the products. In fact, one knows that such products last longer than the observation period. For example, if only 8 out of 10 steel girders would show cracks by the end of the experiment and the total lifetime of all steel girders together would be 7000 days, then a suitable estimate of the parameter is $\hat{\lambda} = \frac{8}{7000}$. The probability that a steel girder will remain free of cracks for at least one year is thus $P(T > 365) = e^{-\hat{\lambda} \cdot 365} \approx 66\%$.

19.3 Lifetime Analysis Under Different Loads

Usually, lifetime experiments do not run under the same conditions as in reality. In so-called 'accelerated lifetime experiments', the products are exposed to significantly higher loads than in reality. Otherwise, one would have to wait too long for a failure to occur. Based on the behavior at the higher loads, one tries to obtain predictions about the behavior at lower loads.

Figure 19.2 shows the logarithmic lifetimes (= number of load changes) up to the break of 31 tension wires that have been subjected to cyclical loads between $s = 200$ and $s = 1100$ MPa (s for 'stress', MPa $= N/mm^2$). It is clear that significantly more censored observations occur with the low loads than with the higher loads. In order to obtain information about the lifetime when the loads are even lower, a model is required that describes how the lifetime $z = T(s)$ depends on the load s. A simple model is given by the relationship proposed by Olin Hanson Basquin in 1910

$$\log(T(s)) \approx \theta_0 - \theta_1 \log(s) \quad \text{with } \theta_1 > 0. \tag{19.1}$$

Fig. 19.2 Estimated Basquin functions assuming the exponential distribution and the Weibull distribution for lifetimes of free-swinging steel under the cyclic loading s

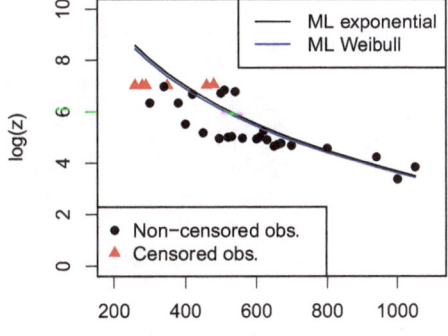

With increasing s, the lifetime becomes shorter. If s approaches infinity, then the logarithmized lifetime $log(T(s))$ approaches $-\infty$, which means for the non-logarithmized lifetime $T(s)$ that it tends towards zero. If, on the other hand, the load tends towards zero, then the lifetime becomes longer and longer.

Back then, Basquin did not have a probabilistic model in mind, but nowadays his model can be combined well with probability distributions and the parameters θ_0 and θ_1 can be estimated using observed lifetime data. For this purpose, a random error is added to Eq. (19.1). This provides a probability distribution of the lifetime as a function of the load. This can, again, be an exponential distribution or the more general Weibull distribution, which is named after the Swedish engineer and mathematician Waloddi Weibull. Figure 19.2 shows the estimated Basquin functions for both random distributions. These are very close together so that one can conclude that the more general Weibull distribution is not necessary in this case. Here, the unknown parameters θ_0 and θ_1 are estimated via the so-called 'maximum likelihood principle'. This means that the parameters are determined so that the probabilities (likelihoods) of the observations are maximized.

19.4 Lifetime Analysis for Products with Several Components

In the case of products made up of several components, the lifetime depends on the lifetime of the individual parts. In such cases, one would like to predict when a certain critical number of components has failed.

Figure 19.3 on the right shows a bundle of tension wires, as used in pre-stressed concrete components of bridges. This bundle comes from an experiment on a pre-stressed concrete girder as shown in Fig. 19.3 (left picture), which contains a total of 35 tension wires for stability. In the experiments carried out, such girders were exposed to cyclic loads from above. Through these loads, the 35 tension wires teared one after the other. After one of these experiments, the bundle of steel wires

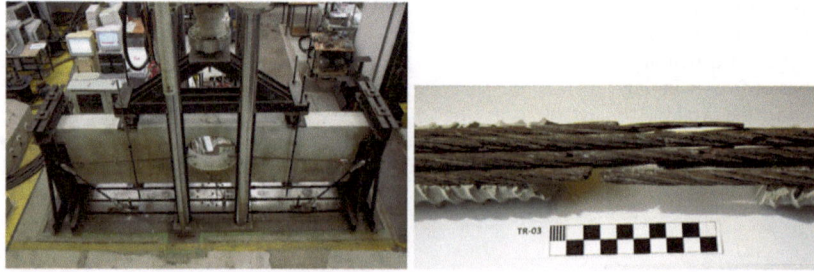

Fig. 19.3 Left: Experiment with a pre-stressed concrete beam at TU Dortmund. Right: Torn tension wires from an experiment as shown on the left (With kind permission of ©Prof. Dr. R. Maurer, TU Dortmund [2019]. All Rights Reserved.)

(Fig. 19.3, right picture) was extracted from the carrier. As one can see in the picture, many wires of the bundle were torn. Here it would be important to know when a critical number of wires has broken so that the overall stability of the girder would be affected.

Such predictions are only possible on the basis of suitable assumptions about the failure of the components. To simplify matters, it is often assumed that the components of the product fail independently of each other and that all components have the same lifetime distribution. However, are such assumptions realistic? The latter assumption only makes sense if the components are of the same type. This is true for the wires in the tension wire bundle in Fig. 19.3 (right picture). The assumption that the components of the product fail independently of each other is, however, obviously not given for the tension wires in Fig. 19.3. The tension wires have the task of sharing an external load. If a single wire fails, the remaining wires have to take on this load. This leads to a so-called load redistribution. In the end, if, in our example, 34 wires have already failed, the last intact wire has to carry the entire load on its own.

19.5   Prediction Intervals

If the intervals between the wire breaks follow an exponential distribution which, via the Basquin model, depends on the external load and the changing internal loads caused by the wire breaks, then forecasts can be made for wire break times at very low external loads on the basis of the observed survival times of the material. Such forecasts are also suitable for external loads that have not been investigated before, because it normally takes a very long time for the first wire breaks to appear. In order to quantify the uncertainty of such forecasts, so-called prediction intervals are calculated. Two different prediction methods are shown in Fig. 19.4 for the forecast of the first four wire breaks with a low external load of 50 MPa (MPa = N/mm^2). These prediction intervals are based on the data in Fig. 19.5, i.e. on the time intervals between wire breaks in ten experiments on concrete beams with external loads between 60 and 455 MPa.

Here, forecasts were calculated for an external load that was lower than the previously used external loads. The vertical red line in Fig. 19.4 shows the time of the actual first wire break, which, indeed, lies in the prediction interval for the first wire break. The tip of the red arrow indicates the end of the experiment, which, at that time, had already been running for almost six months. It therefore remained open, whether the further wire breaks fall within the calculated prediction intervals. However, it was already a success that the first wire break was correctly determined with the statistical methods.

Still, one important effect is left out, namely damage accumulation. In the model used above, the risk of a further failure depends on the external load and the number

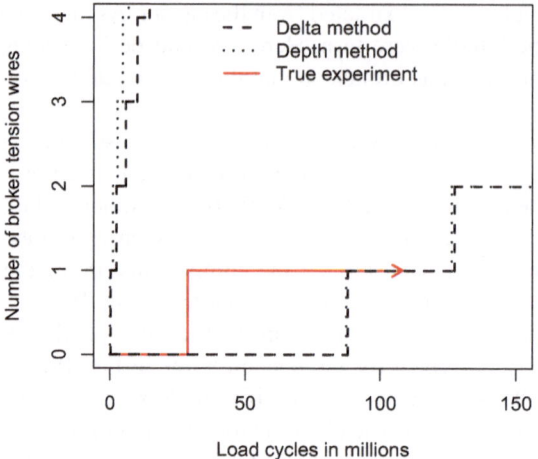

Fig. 19.4 Prediction intervals (left and right limits) from two different prediction methods for the number of load cycles up to the first, second, third, and fourth wire break at 50 MPa external load based on the results in Fig. 19.5

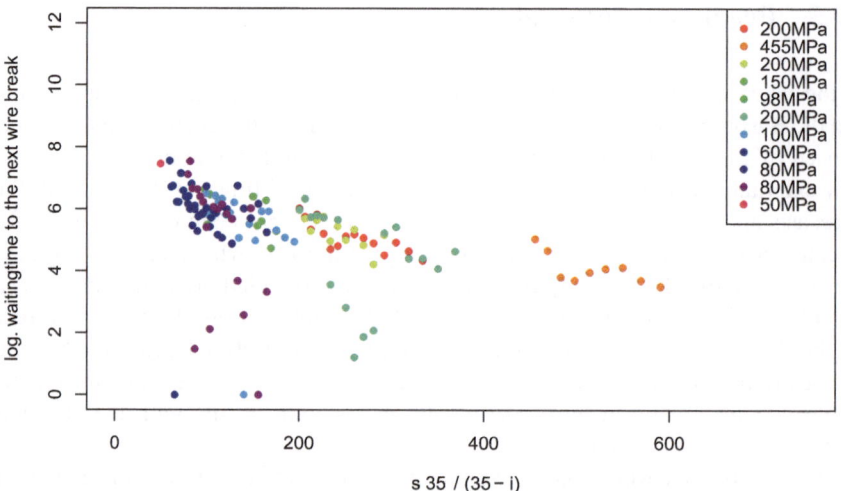

Fig. 19.5 Waiting times measured in 'number of load cycles between successive wire breaks' depending on $\frac{s}{35-i}$ for 11 experiments with external loads of $s =$ 50, 60, 80, 98, 100, 150, 200, 455 MPa and $i = 0, 1, \ldots$ already torn tension wires of a maximum of 35

of previous wire breaks; it does not take into account how long it took for previous wires to break and when exactly the last failure occurred. This is unrealistic. Because the longer a wire is exposed to a certain load, the greater the risk that it will break. Current research tries to integrate such damage accumulation into a suitable model and to develop more realistic forecast intervals for future wire breaks.

19.6   Outlook

In the case of products made up of different components that also interact differently with one another, a prediction of the remaining service life becomes even more complicated. With suitable model assumptions, however, predictions can sometimes be made. As much information as possible is to be used for this purpose. If the failure is caused by cracks, it makes sense to model the crack growth. One can start with the formation and growth of micro-cracks, as shown in Fig. 19.1. However, many questions remain open and there is still much to be done to fully understand the processes of aging of technical products.

19.7   Further Reading

The experiments for this chapter were carried out at TU Dortmund. One of the earliest works on the relationship between service life and load is Basquin, O.H. (1910): "The exponential law of endurance tests," American Society of Testing Materials 10, 625–630. More recent work on this topic at TU Dortmund University includes Heinrich, J., Maurer, R. (2018): "Check of resistance to fatigue on existing prestressed concrete bridges by monitoring," Proceedings of the 12th fib International PhD Symposium in Civil Engineering, 987–994, or Leckey, K., Müller, C. H., Szugat, S., Maurer, R. (2020): "Prediction intervals for load sharing systems in accelerated life testing," Quality and Reliability Engineering International 36, 1895–1915.

Chapter 20
Durable Machine Components: How Statistical Design of Experiments Optimizes Wear Protection

Sonja Kuhnt, Wolfgang Tillmann, Alexander Brinkhoff, and Eva-Christina Becker-Emden

Abstract The service life of machine components is significantly extended by wear protection coatings. Statistical design of experiments and data analysis can be used to efficiently improve the quality of protective coatings.

20.1 Wear Protection Through Coating

Industrially manufactured products must meet high quality requirements. This also includes wear protection, through which machine parts are provided with an additional wear protection coating. The coating process must ensure the required thickness and hardness of the coating, while at the same time keeping porosity as low as possible. A high deposition efficiency also ensures that the coating entails as much of the actual material to be applied as possible. The required measurement of porosity, deposition efficiency, and coating thickness and hardness cannot be carried out during the production process but only under laboratory conditions, more precisely in the process of destroying the coating. At the same time, environmental influences that cannot be controlled for as well as wear processes of the coating device may lead to changes in the coating quality, even in a well-adjusted coating process. In cooperation between mechanical engineering and statistics, we develop methods to adjust the production process in an optimal way, i.e. on a daily basis and

S. Kuhnt (✉) · E.-C. Becker-Emden
FH Dortmund, Faculty of Computer Science, Dortmund, Germany
e-mail: sonja.kuhnt@fh-dortmund.de; eva-christina.becker-emden@fh-dortmund.de

W. Tillmann
TU Dortmund, Department of Mechanical Engineering, Dortmund, Germany
e-mail: wolfgang.tillmann@udo.edu

A. Brinkhoff
TU Dortmund, Dortmund, Germany
e-mail: alexander.brinkhoff@tu-dortmund.de

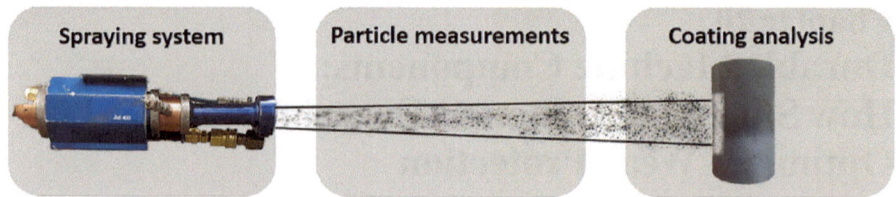

Fig. 20.1 Thermal spray process. Metal powder is heated up in the spray gun, accelerated, and deposited on the work piece by means of a hot gas jet

without destructive component testing. For this purpose, we use measurements of the properties of the sprayed particles during the application process.

We focus on the process of thermal spraying, i.e. the application of a metallic or non-metallic coating to a surface. In this process, the heated material is sprayed onto the surfaces to be coated. When the spray particles hit the surface, the coating forms successively (see Fig. 20.1). Coating processes are classified according to the type of energy source used to spray the particles. For example, the flame spraying process, where the energy comes from combustion, is different from the arc spraying process, where the energy is provided by an electric arc.

For our research, we work with the high velocity oxygen fuel process (HVOF process), in which kerosene is used as liquid fuel. This process is a further development of the flame spraying process, in which the hot gas jet is additionally accelerated and thus the particles reach a very high velocity. The HVOF process depends on a number of factors, which, in the right combination, produce an optimal coating. In addition to the flows of the various fuels, gases, and powder supply, the configuration parameters of the spray torch are also important for achieving the most consistent coating quality possible. In our study, we investigated the influence of the following parameters: the distance from the spray gun to the component (spray distance), the amount of kerosene, the ratio of oxygen to fuel, and the powder feed rate. Drawing on available expertise in the area of thermal spraying as well as on results from preliminary tests, ranges were defined within which the values of such process variables could be varied in a reasonable way. Statistical methods then helped to find settings of the four process factors involved that produce the best possible coatings.

20.2 Optimization Through Statistical Design of Experiments

Intuitively, trying out promising factor settings based on expertise one after another seems to be a sensible experimental approach. However, each experimental run requires effort and time, so that usually only a limited number of runs can be carried out. Therefore, options of possible settings are often searched very selectively, so that suitable settings that could not be inferred from prior knowledge may be neglected.

Statistics proceeds differently. A plan (experimental design) is set out in advance as to how many experimental runs will be carried out and with which settings. These settings are chosen in such a way that as much information as possible is obtained by means of as few experimental runs as possible. Therefore, statistical experimental designs can be quite different, depending on the aim of the study.

As a first step, we limit our question to a technical process including one quality characteristic (target variable), which depends on the settings of two factors (influencing variables) A and B and inquire into the question of how to find settings for which the quality characteristic assumes the highest possible values. In this respect, the first step is to determine different levels, i.e. specific settings of the factors, drawing on knowledge from both statistics and mechanical engineering. For a full factorial experimental design, all possible combinations of these settings are subsequently considered. If only one high and one low setting is specified for each factor, comparisons of the mean values of the measured values for the quality characteristic between the high and the low settings provide information on how large the effect of the factor on the quality characteristic is. Often it is sufficient to study only a well-selected proportion of all possible combinations of factor levels in experiments. These fractional factorial designs are extremely helpful, especially for experiments including a high number of factors, so that otherwise the number of experimental runs would exceed time and cost constraints.

In search of the best possible settings of the influencing factors, the factors have to be varied on more than two levels. Important examples of such experimental designs are 'central composite designs' (CCDs) since they have proven to be very useful in industrial applications. These designs also contain points in the center of the experimental space as well as so-called star points on the coordinate axes. Figure 20.2 shows the positions of the measurement points in the experimental space. An experimental design contains the setting for one experimental run in each row. Figure 20.2 shows so-called coded setting values for factors A and B in the experimental design. In order to reveal the general structure of the design, the high setting value is transformed to 1 and the low to -1. The coded value 0 corresponds to the setting right in the middle between the high and low values on the natural scale of the factors. In our design, the star points lie outside the range between the low and high values. To generate these design points, the distances of the low and the high values to the midpoint of their range are multiplied by the value $\sqrt{2}$. This leads to the codings $-\sqrt{2}$ and $\sqrt{2}$. In general, the star points can also lie between the high and low settings or can even be identical to these settings. The decision on the choice of the star points depends not only on considerations specific to mechanical engineering (e.g.: Are points with values lower than the low or higher than the high setting sensible?) but also on statistical-theoretical considerations.

Through 'repetitions', it is possible to estimate to what extent experimental results vary solely due to unavoidable external influences and measurement inaccuracies. These unavoidable, random scatterings are to be separated from differences in measurement results due to actual effects. Besides repetitions, 'randomization' is

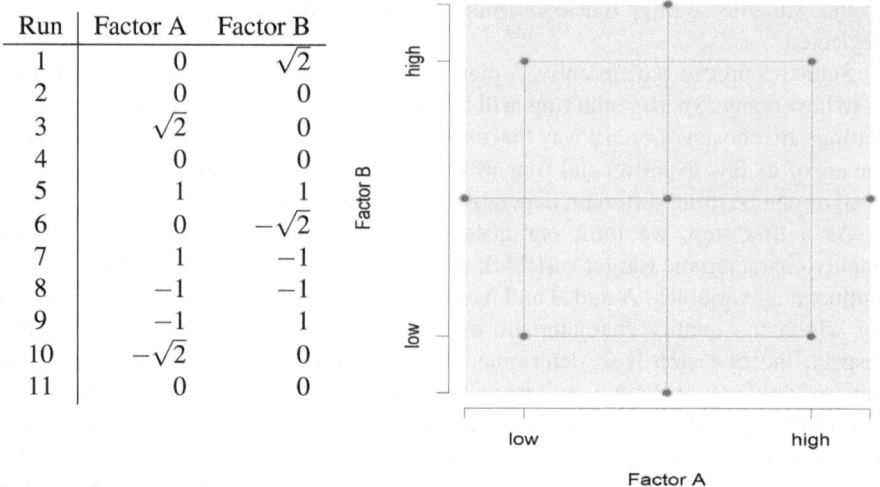

Run	Factor A	Factor B
1	0	$\sqrt{2}$
2	0	0
3	$\sqrt{2}$	0
4	0	0
5	1	1
6	0	$-\sqrt{2}$
7	1	-1
8	-1	-1
9	-1	1
10	$-\sqrt{2}$	0
11	0	0

Fig. 20.2 Randomized experimental setup with coded factor values (left) and graphical representation of the design points (right) for a central composite design for two factors A, B with external star points

another important tool in experimental design to deal with unavoidable confounding effects. This involves randomly selecting the order of experimental runs. Figure 20.2 contains a possible randomization of the experimental design. Repetitions are included in the form of three center points, with all settings at the center level 0 in each case.

After all experiments have been run, 11 measured values of the target variable are available. In the left part of Fig. 20.3, the measured values are plotted over the

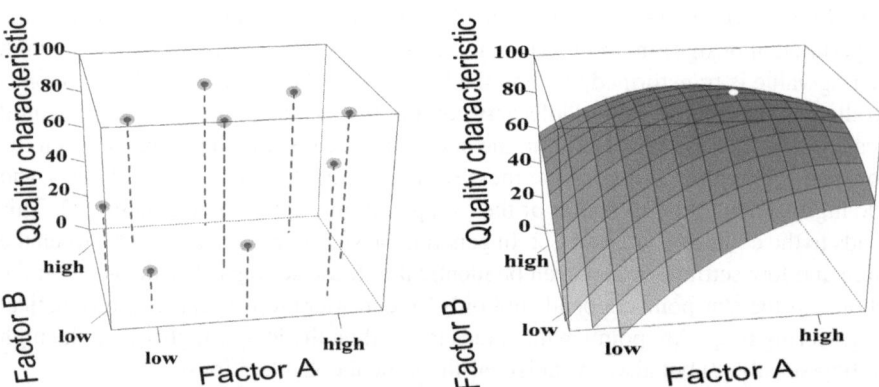

Fig. 20.3 Graphical representation of the measured values of the quality characteristic (target variable) from the 11 runs of the experimental design (left) and the prediction model as a response surface through the measurement points (right), around which the measured values scatter. The largest predicted value for the quality characteristic is marked by a white dot

experimental space. Within the performed experiments, the largest value (82.07) was observed for the setting $A = \sqrt{2}$ and $B = 0$.

This is where statistical modeling comes in, in order to gain information beyond the pure measured values and to find even better settings. By means of a mathematical-statistical model that takes into account non-controllable, random fluctuations of the measured values, the relationship between factors and quality characteristics is described and explored. The surface plot in the right part of Fig. 20.3 shows a fitted regression model that runs through the observed values. This model allows the prediction of values of the quality feature for unobserved points in the experimental space. For example, it can be seen that with higher values of factor A, the quality generally increases. The same is true for factor B. However, in both cases, this depends on the interaction with the other factor. As the right part of Fig. 20.3 illustrates, for low values of factor A, the quality of the model is generally low, but it increases with increasing values of factor B. As the same figure illustrates, if the values A are very high, so is the model quality, everywhere. From the lowest to the highest value of B, first a slight increase and then a slight decrease of quality can be seen. Apparently, so-called 'interaction effects' exist here. If we now take a specific setting of the factors, say $A = 0.2$ and $B = 0.2$, observations at this point are predicted to have a mean value of 84.96. The highest expected value of the quality characteristic is estimated to be 87.89 at the setting of $A = 0.52$, $B = 0.50$.

20.3 Challenges in Real Coating Processes

The analysis of real thermal spray processes is much more complex because of the large number of quality criteria. Porosity, deposition efficiency, coating thickness as well as coating hardness have to meet certain requirements. If these cannot be achieved simultaneously, possible compromises have to be found. The starting point for simultaneous optimization of several quality characteristics is again provided by prediction models obtained from experimental data.

Figure 20.4 visualizes models for the two quality characteristics 'coating hardness' and 'coating thickness' modeled by means of the four influencing variables kerosene, ratio of oxygen to fuel, spray distance, and powder feed rate, which were varied according to a CCD based on the coded values -2 and 2 for the star points.

The coating hardness is displayed in the upper surface plots and the coating thickness in the lower ones. The amount of kerosene (K) and the powder feed rate (FDV) vary across the x- and y-axes. The ratio of oxygen to fuel is held at the medium setting value in all plots (coded as 0), the spray distance changes from low (-1) to high (1) from left to right.

Let us first look at the coating hardness. The hardness increases with higher kerosene levels (upper surface plots in Fig. 20.4). More kerosene increases the pressure in the combustion chamber and thus the kinetic energy transferred. The coating becomes harder. The effect of the powder feed rate is not that clear. It has an influence on how well the powder melting process works. With a low amount of

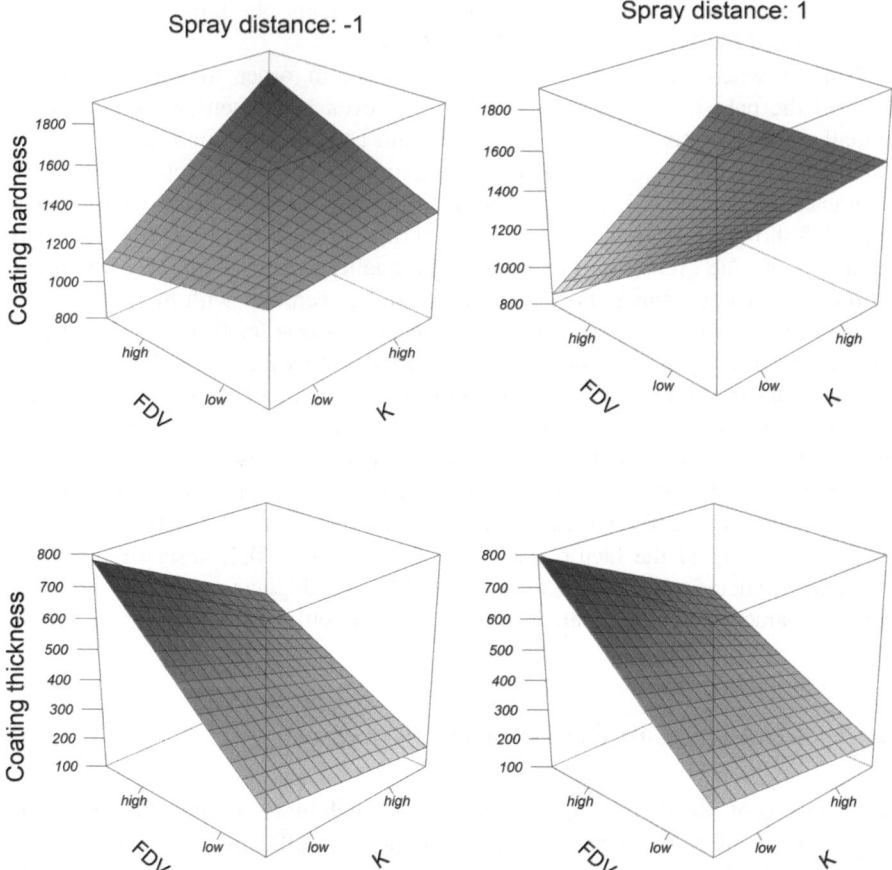

Fig. 20.4 Statistical prediction models for coating hardness and coating thickness as a function of three influencing variables. For the surface plots shown, the ratio of oxygen to fuel is fixed at the medium setting

kerosene, a higher powder feed rate leads to a lower hardness, whereas with a high amount of kerosene, it leads to greater hardness. Therefore, an interaction between the two factors can be observed here.

Another interaction exists between the powder feed rate and the spraying distance. For a smaller spraying distance, higher temperatures are achieved when the coating hits the work piece, which leads to an easier processing of large powder quantities. This results in more advantageous microstructures on the surface and harder coatings. The surface plots for the coating thickness clearly show that a higher powder feed rate leads to thicker coatings (lower surface plots in Fig. 20.4), as more powder reaches the work piece in the experiment. A higher spray distance results in a lower particle velocity. Slower particles bounce off the surface less

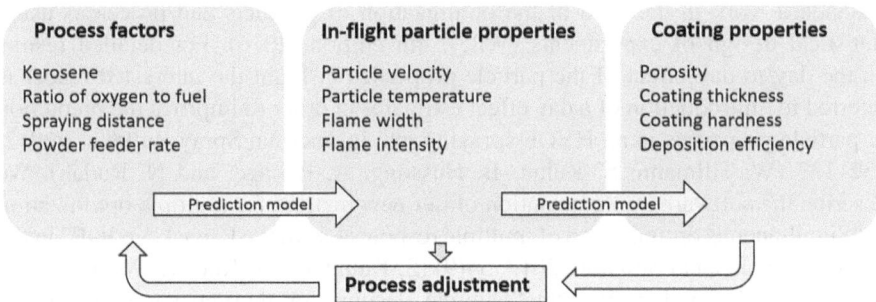

Fig. 20.5 Modeling of the thermal spraying process

frequently. Therefore, the prediction area for the coating thickness slightly moves upwards at a higher spray distance.

The highest overall coating hardness is predicted at a very high powder feed rate and a very high kerosene flow rate (highest point in the upper left plot of Fig. 20.4). Thick coatings, however, are produced at distinctly different settings (lower plots of Fig. 20.4). Since we want to identify the one setting to be used in practice, a compromise setting must be found. Such a compromise setting would require that no other settings exist at which one quality characteristic improves without corrupting another.

Often, an additional adjustment of the best possible setting identified in the experiments is necessary due to non-controllable daily environmental effects. In the process of spraying, these mainly relate to properties of the particles during flight. The importance of the particle temperature and particle velocity for the quality of the machine components has already been discussed. However, the width and intensity of the combustion flame are also important. For these four characteristics, predictive statistical models can be developed on the basis of the four process factors (see Fig. 20.5). This way, we get two-step models for the prediction of the quality characteristics as a function of process factors and (predicted) in-flight particle properties (Fig. 20.5). By repeatedly measuring the particle properties in flight, the spray process can be continuously adjusted on a day-to-day basis. This way, we achieve a day-to-day optimization and control of the thermal spray process, taking into account process uncertainties.

20.4   Further Reading

Statistical design of experiments has been used successfully for a long time, first for agricultural experiments and later in industrial development. The monograph "Response Surface Methodology – Process and Product Optimization Using Designed Experiments" by R. Myers, D. Montgomery, and Ch. Anderson-Cook is a standard work in the field of the optimization of products and processes using statistical design of experiments (Wiley, 4th edition, 2016). For detailed results on the day-to-day effect of the particle properties in flight the interested reader is referred to "Introduction of a day effect estimator in order to improve the prediction of particle properties in an HVOF spraying jet" in Thermal Spray Bulletin 2/2012, 132–133 (W. Tillmann, S. Kuhnt, B. Hussong, A. Rehage, and N. Rudak). We describe the software implementation of our new method for multiple optimization in "Simultaneous optimisation of multiple responses with the R-package JOP" in the Journal of Statistical Software 2013, 54(9) (S. Kuhnt and N. Rudak). We developed special experimental designs in "Optimal designs for thermal spraying" in the Journal of the Royal Statistical Society, Series C, 2017, 66(1), 53–72 (H. Dette, L. Hoyden, S. Kuhnt, and K. Schorning). More recently, we also looked at internal diameter coatings in "Effect of the Spray Parameters on the Particle Behavior and the Coating Properties During ID Warm Spraying of Fine WC-12Co Powders $(-10 + 2\mu m)$" (W. Tillmann, I. Baumann, A. Brinkhoff, S. Kuhnt, E.-C. Becker-Emden, and A. Kalka, Thermal Spray 2021: Proceedings from the International Thermal Spray Conference Virtual. ASM, 283–289).

Part V
Intricacies of Measurement

Chapter 21
Measuring the Immeasurable: Statistics, Intelligence, and Education

Philipp Doebler, Gesa Brunn, and Fritjof Freise

Abstract Educational concepts such as mathematical proficiency have their statistical equivalent in so-called latent variables, i.e., variables which cannot be
observed directly. To gain insights into learning processes, the same latent variable
is measured repeatedly, resulting in a so-called 'latent learning trajectory'. Often,
the learning trajectory of an average student is of interest, say to understand effects
of programs on a group of students. However, a student's deviation from the mean
trajectory is also important to quantify to assign individualized measures. The
present chapter provides details on a flexible statistical model for the description
of general increases or decreases as well as individual deviations.

21.1 Educational Tests and Education

Since the OECD published their report on the large-scale PISA study in the year
2000, which has revealed serious deficits in the education of German students,
German media love to refer to the initial public reactions to the report as the 'PISA
shock'. Germany collectively wondered whether German students are dumber
than their Finnish or Korean peers. However, how can we measure intelligence,
education, and educational progress in objective ways? Even though scores from
an achievement test do not necessarily correspond to education in general, they are

P. Doebler (✉)
TU Dortmund, Department of Statistics, Dortmund, Germany
e-mail: doebler@statistik.tu-dortmund.de

G. Brunn
Deutscher Wetterdienst, Research and Development, Offenbach, Germany
e-mail: gesa-marie.brunn@dwd.de

F. Freise
University of Veterinary Medicine Hannover, Department for Biometry, Epidemiology and
Information Processing, Hannover, Germany
e-mail: fritjof.freise@tiho-hannover.de

C. Weihs et al. (eds.), *Statistics Today*, Society, Environment and Statistics,
https://doi.org/10.1007/978-3-662-68907-3_21

often the starting point of research into problems of educational systems and they are also the basis for personalized educational interventions. Statistics analyzes and models test results and helps to gain robust insights into academic progress.

In general, learning progress is difficult to measure: Unlike variables such as height or body weight, education is a 'latent variable', i.e., a variable that cannot be observed directly. Answers to test items, i.e. all kinds of test questions in psychological and educational tests, are 'indicators' at best, by no means to be mistaken for answers to the actual concepts being measured.

One statistical way to describe the relationship between test scores and latent abilities is 'item response theory'. In the following, we specifically look at longitudinal surveys on the educational achievement of study groups (e.g., classrooms) or individual students. Nowadays, such surveys are often computer based and show whether study groups or individual students respond to educational measures. Test designers create items which try to combine different aspects of a broad concept like 'reading' or 'maths' in such a way that difficulty and content of the items are appropriately balanced. This is to avoid, for example, ending up with difficult geometry tasks only, without taking into account items of practical or everyday relevance.

Even when test construction is relatively straightforward, test designers still have to deal with a computerized implementation, e.g., integrating a test into an engaging 'serious game' played for training, or providing real-time feedback to educators in a web-based application. One example is the rather simple reading test from the 'Levumi' online test system.[1] One of the tasks of the test is reading aloud syllables. Levumi test items correspond to individual syllables (such as *fa, la,* or *te*). Under time pressure, children are asked to read randomly selected syllables. A response to a syllable item is correct if it is read correctly within the allotted time. Why is this basic task informative of reading proficiency, the skill we use to gain deep insights into texts? Sure enough, this test is only a small part of a bigger picture of reading proficiency. However, reading syllables fluently—with speed and accuracy—is a prerequisite for fast word recognition and for comprehension of whole sentences. As a consequence, diagnosing deficits in the automaticity of syllable reading is helpful, especially when addressing learning disabilities.

21.2 Latent Variables and Their Indicators

Many latent ability models rely on the key assumption that a single number per person is sufficient to capture their ability in the specific task at hand. This unidimensionality of the latent variable is a helpful simplification of a potentially

---

[1] Cf. Gebhardt, Diehl & Mühling (2016): Online-Lernverlaufsmessung für alle Schülerinnen und Schüler in inklusiven Klassen. Zeitschrift für Heilpädagogik, 10, 444–453.

more complex real world and has proven useful in practice. Scales of latent variables are arbitrary; in fact, one has to deal with them a little to understand their units.

For example, Ernest, a fifteen year-old ninth grader, might achieve -0.7 as a maths score on a continuous scale centered at zero, which is assumed to be the value of an average student. It is important to note here that Ernest just never does his homework, which is why his score is below average. Esther, however, who is a diligent student, might achieve a 1.8. Thus, Ernest's and Esther's scores are $2.5 = 1.8 - (-0.7)$ points apart. But what exactly does a difference of 2.5 mean for our investigation? To clarify this, mathematicians typically define what a difference of 1 means. One convention is to interpret 1 as the standard deviation of the population. This means that Esther is 2.5 standard deviations better than Ernest (which would, in fact, be a big difference, amounting to several years of schooling Ernest would need to catch up). In other models, the scale of the latent variables is realized in so-called 'logits'. This is defined as the natural logarithm (*log*) of the ratio of the probability of scoring 1 and the probability of scoring 0. A score of 1 means that the syllable was read correctly in a pre-fixed time frame and 0 means that a student was not able to produce a specific syllable within the envisaged time frame or made a mistake. If Ernest's probability for reading the syllable *vi* correctly is 33%, we say that Ernest's chance for being correct is 1 to 2, meaning that in 3 trials he would produce 1 time the correct syllable and 2 times not. The calculated difference of 2.5 between Esther and Ernest would then mean that Esther's chance of reading the syllable correctly in the aspired time frame would be higher by a factor of $e^{2.5}$, i.e. it would be $e^{2.5}$ to 2, amounting to a probability of 85.9% $(= e^{2.5}/(e^{2.5} + 2))$ to read correctly in the time frame.

Figure 21.1 visualizes the idea of measuring competence on a continuous scale and also introduces the concept of item difficulty. The items here represent the student competence in fractional arithmetics. The scale also illustrates and compares Ernest's and Esther's competencies. We instantly see that Esther has a better latent ability. In this example, as in many models, the items are on the very same scale as the students. While items do not exhibit a latent competence, their points on the continuum correspond to latent item difficulty. The further to the right an item is, the more difficult it is, i.e., large values indicate high difficulty. Why is it reasonable to have items and students on the same scale? It helps to illustrate and compare student competence in relation to item difficulty, e.g. to express that the difference in the competence of Ernest and Esther is similar to the difference in the difficulty of item 1 and item 2.

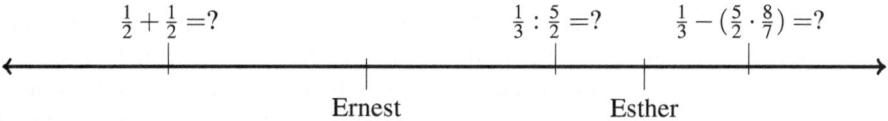

Fig. 21.1 Competence in fractional arithmetics reduced to a single number as in the models discussed here. Items and persons are on the same scale

Fig. 21.2 S-shaped item characteristic curve which provides the probability of a correct response as a function of the latent variable (latent ability). Here the item difficulty is assumed to be $\beta = 0$

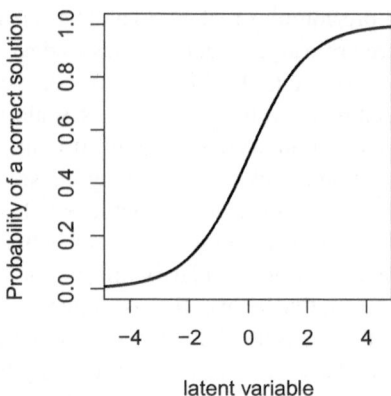

Statistical models for correct and incorrect answers are not deterministic but probabilistic: They specify probabilities for each potential combination of correct and incorrect answers. Figure 21.2 shows a typical relationship between a latent variable θ and the probability p of a correct response. The S-shaped curve in Fig. 21.2 for the probability p as a function of the latent ability θ can be stated as follows:

$$p(\theta) = \frac{\exp(\theta - \beta)}{1 + \exp(\theta - \beta)}$$

Functions like this, in which p is a function of θ, are called 'item characteristic curves'. The specific item characteristic curve in Fig. 21.2 reflects the assumption that as the student's ability increases, so does the probability of the student providing the correct solution. For example, if the item is scored 0 for an incorrect answer and 1 for a correct response, Esther, with her ability of 1.8, would score an average of 0.84 points on an item with a difficulty of 0, for which the item characteristic curve is shown in Fig. 21.2. In other words, if we did not know yet whether Esther would be able to respond correctly to the item in Fig. 21.2 but if we knew Esther's ability (= 1.8) and the item's difficulty (= 0), we would predict a 84% chance ($p = 0.84$) that Esther would be successful.

Item characteristic curves even allow indirect comparisons between subjects who were presented with different items. Based on item characteristic curves, we can calculate probabilities and infer latent ability from observed response patterns. If the statistical model fits the observed responses well, we can exchange some or even all items and could still estimate the latent ability since the curve is assumed to be generally valid. This is important for at least two reasons: First, students do not like to take the same test again. Especially in longitudinal surveys, it is unfavorable to use the same item several times within short intervals, not only because it might frustrate or bore students, but also because stochastic dependencies would have to be taken into account because students might memorize certain parts of the earlier test. Second, we can give a basic test to Ernest and an advanced test to Esther, with both neither overchallenged by the test material nor bored. Thus, Esther and Ernest may work on completely different items, but, thanks to statistics, the estimates of the latent variables can be compared on the same scale.

21.3 A Statistical Model for Learning Progress Diagnostics

Figure 21.3 illustrates the model which we have developed for estimating the learning trajectories in the Levumi online platform. The model assumes a continuous but otherwise arbitrary, general mean trajectory of learning. Traditional models use

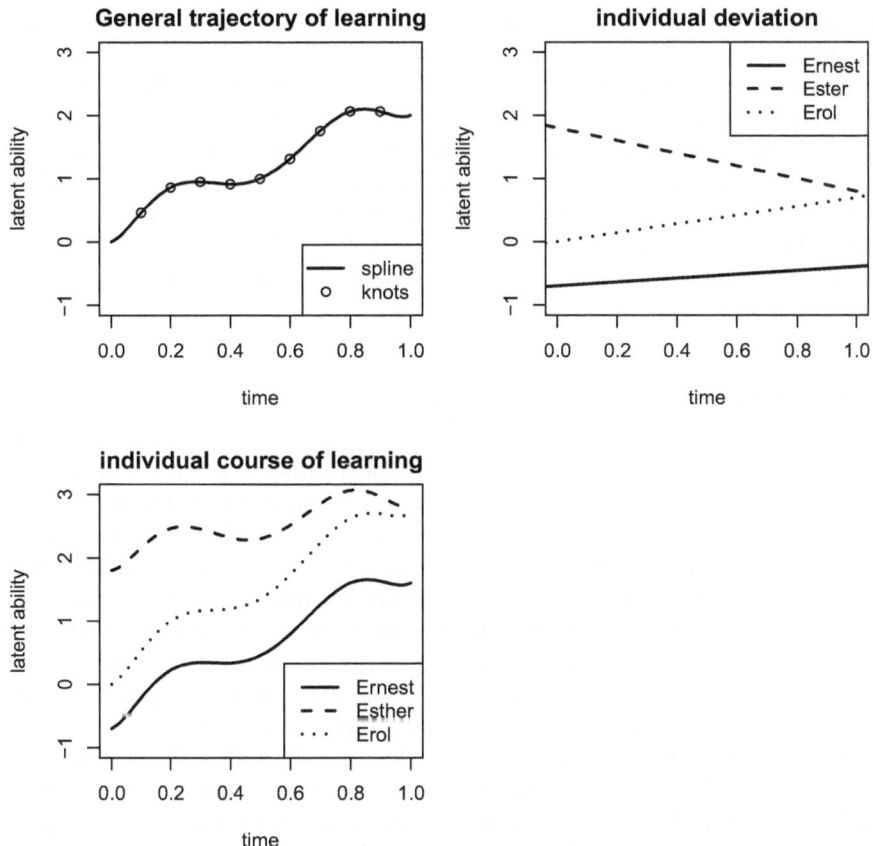

Fig. 21.3 The general trajectory (top left) is modeled with a spline, which is given by a third-degree polynomial between each two grid points. The general trajectory is combined with deviations of individual children from the mean course, which are assumed to be linear (top right). By adding the general trajectory and the individual deviation the individual courses of the latent abilities result (bottom left)

linear functions for mean learning trajectories, but this might not fit empirically. We set up our model so that the general trajectory is deliberately very flexible and reflects that increases in performance might be irregular, e.g., increases could be particularly strong after relevant material has been covered in class. However, decreases are also possible, for example after the winter vacations.

Mathematically, we can describe a flexible trajectory by means of so-called 'splines', which are piecewise polynomials that are smoothly connected to each other (for an example, cf. Fig. 21.3, top left). Piecewise polynomials refer to the following idea: If one zooms in on the spline, a small section is a low degree polynomial (e.g., third degree). The polynomial sections are delineated by so-called knots. Many functions are locally well approximated by polynomials, but the flexibility of the spline comes at a price. We have to estimate the coefficients of many polynomials. However, because of the large number of classrooms that have already used the Levumi system and have delivered observations, we have enough data to estimate the coefficients of the spline.

In addition, the model makes a prediction for each child's latent ability in reading at each point in time t. These 'individual trajectories' (cf. Fig. 21.3, bottom left) are the sum of the general trajectory and the individual deviations from it (cf. Fig. 21.3, top right). Together with the item characteristic functions such as the one in Fig. 21.2, our model predicts probabilities to successfully solve the task for each item and each child at all points in time. At the level of individual students, it becomes possible to estimate an individual progression after several measurements.

21.4   From Data to Latent Variables

Figure 21.4 shows the times at which the syllable reading test results were fed into the Levumi servers. Since the syllable reading test was conducted with individual students, it was difficult and very time-consuming to collect data from many students at the same measurement point. School vacations (here those of the German federal state of North Rhine-Westphalia) interrupted the data collection. Table 21.1 shows data as they are available after one measurement at one time point, here highly simplified for only three children and eight test items. As usual, 1 means that the syllable was read correctly. If a student is not able to produce a specific syllable within the envisaged time frame or makes a mistake, the result is 0.

However, the response pattern for Esther does not tell us directly what her latent ability is. Our intuition tells us, that the more responses are correct, the higher the latent ability, but where is Esther's point in the ability continuum? Assuming that item difficulties are known with sufficient precision from previous studies, Esther's current ability can be estimated by means of the method of maximum likelihood (cf. Fig. 21.5, top left). The likelihood ('plausibility') of a person's ability is the model-implied probability of the observed response behavior as a function of the ability. In Fig. 21.5, top left, the likelihood is shown as a function of Esther's current latent

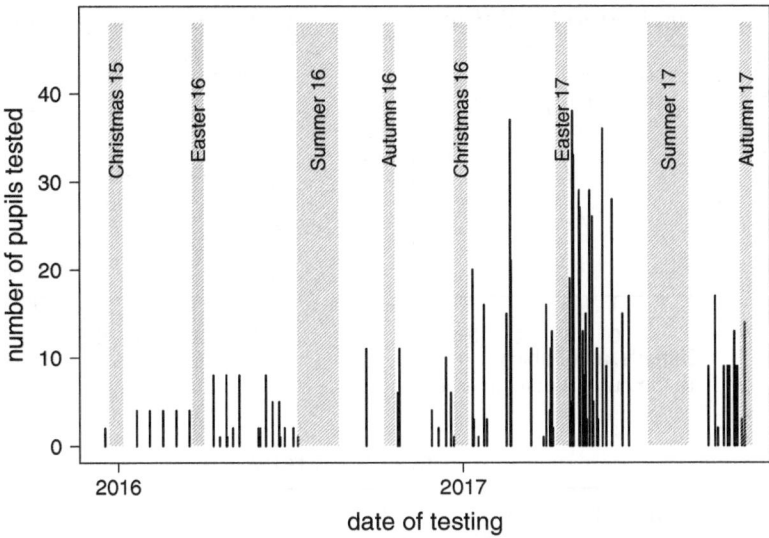

Fig. 21.4 Dates of Levumi tests. School vacations (here those of the German federal state of North Rhine-Westphalia) interrupted the data collection

Table 21.1 Example of test data

	Item index							
Child	1	2	3	4	5	6	7	8
Ernest	1	0	0	1	0	0	0	0
Esther	1	1	1	1	0	1	1	1
Erol	1	1	1	0	0	0	0	1

ability and the maximum (here approximately at 2) is used as an estimator for her ability.

Just as important as describing individual learning progression is to check whether it lags behind a reference trajectory. In Fig. 21.5, Ernest's trajectory can be seen in the upper right-hand corner. The reference trajectory is not necessarily the same as the average trajectory, because for a student with a weak starting level, a reasonable learning goal could be to develop parallel to the average trajectory— not falling behind further. If we want to know whether Ernest's learning curve is parallel to the general trajectory, we can calculate the likelihood ratio, i.e., the ratio of the likelihood at the maximum and the likelihood of average growth, which is illustrated in the lower part of Fig. 21.5 for Ernest. If the ratio is very large, a parallel course is unlikely. The average increase is 2.00 here, and Ernest's increase with maximum likelihood is 2.41. Figure 21.5, bottom, also shows that the ratio of Ernest's maximum likelihood and the likelihood at the average increase is 4.74. From this, we can conclude that Ernest has an above average increase, albeit from a sub-par starting level.

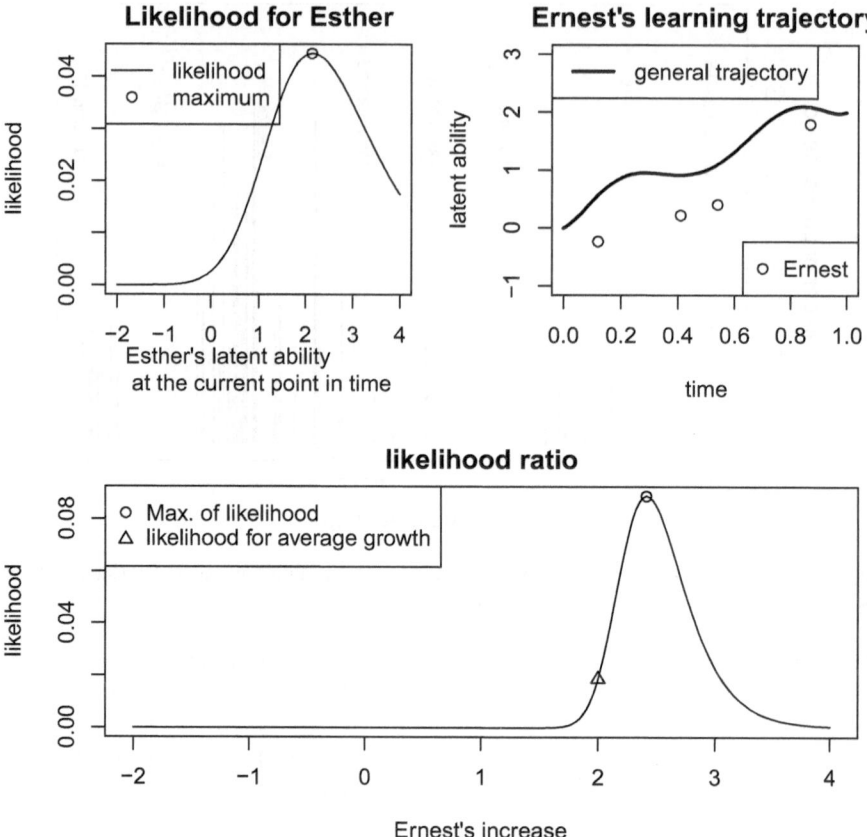

Fig. 21.5 In the top left part of the figure, the likelihood is shown as a function of Esther's current latent ability. At the top right, several measurements of Ernest's latent ability are illustrated, whose performance is below the general trajectory but catches up over time. Time is scaled from 0 (earliest start of observations) to 1 (latest end of observations). On the lower panel's x-axis, potential values of Ernest's increase from start to end are varied, on the y-axis the corresponding likelihoods are shown

Any inference from the data to the associated latent abilities is subject to error, due to the indirect nature of making inferences or the probabilistic nature of modeling. In addition to the estimate of Esther's current ability, an estimate of the associated uncertainty of that estimate, called the 'standard error', can be calculated. The higher the standard error, the less reliable the test. Practically, a compromise between the time spent testing (instead of teaching!) and a reliable enough latent ability estimate needs to be found.

21.5 Further Reading

For many more details on educational testing, see the special issue of the Journal of Educational Research Online on "Design, Construction and Analysis of Progress Monitoring Assessments in Schools" from the year 2022, in particular Brunn, G.; Freise, F.; Doebler, P., "Modeling a smooth course of learning and testing individual deviations from a global course," 89–121—DOI: 10.25656/01:24874. The authors would like to thank Markus Gebhardt for providing the Levumi data on syllable reading.

Chapter 22
Uncovering Embarrassing Truths Through Statistics

Andreas Quatember

Abstract Surveys on sensitive topics often suffer from non-response and untruthful answers. Fortunately, statistics offers methods to elicit even embarrassing truths from respondents.

22.1 The Method of Indirect Questioning

In 1965, the American statistician Stanley L. Warner had a brilliant idea: If you want to find out how large the proportion of students is who cheat on exams (or of notorious fare dodgers or of people who never brush their teeth), then you should not directly ask them about their behavior. With such sensitive topics, even in anonymous surveys, many people do not answer at all or deliberately provide incorrect answers, probably due to shame. However, Warner, in a groundbreaking paper, showed how honest answers on sensitive topics can still be elicited: Do not present students with just one question, but two:

1. "Do you belong to the group of those students who did knowingly violate the exam rules?"
   and
2. "Do you belong to the group of those students who did not knowingly violate the exam rules?"

When these questions are presented, the students are asked to use, hidden from the interviewer, a randomization device which selects one of these questions (like a wheel of fortune stopping at one of two possible outcomes) at predetermined probabilities. Depending on the outcome, either question 1 or question 2 has to be answered. Since no one other than the respondent knows which question he or she actually had to answer, there is no incentive to lie or not answer the question at all

A. Quatember (✉)
JKU Linz, Institute of Applied Statistics, Linz, Austria
e-mail: andreas.quatember@jku.at

C. Weihs et al. (eds.), *Statistics Today*, Society, Environment and Statistics,
https://doi.org/10.1007/978-3-662-68907-3_22

because the privacy of the respondent remains protected. But what does this imply with respect to the calculation of the proportion of cheaters of interest in the present study? Two pieces of information are available in this respect: The proportion of 'yes'-answers in the survey and the known probabilities of having to answer the 1st or 2nd question utilized by the randomization device. Let us assume, for example, that the experiment includes a group of 100 students and that the randomization device has been set in such a way that question 1 has to be answered with a probability of 0.8 and question 2 with a probability of 0.2. On the one hand, if all 100 students indeed cheated, only the expected 80 of them who received question 1 would answer 'yes'. On the other hand, if no one cheated, we would expect only 20 (those with question 2) to answer 'yes'. What if half of all students actually violated the exam rules? Then we would assume that half of the expected 80 who received question 1 would answer 'yes' and also half of the expected 20 with question 2 would do so. This would then result in an expected number of 50 'yes'-responses. An estimator for the desired cheating proportion is thus obtained by the ratio of two differences, namely the difference of the number of actual 'yes'-answers to the minimally expected 20 'yes'-answers and the difference of the maximum number of 'yes'-answers of 80 to this expected minimum. With 35 given 'yes'-answers among 100 students, for example, this would result in an estimated cheating proportion of

$$(35 - 20) : (80 - 20) = 0.25 = 25\%.$$

22.2 A Modification of the Original Idea

In the context of an advanced course in statistics at the Johannes Kepler University (JKU) Linz, Austria, the author of the present chapter conducted a similar survey experiment on students' violation of examination rules, in this case the violation of rules for the basic course of the previous semester. Unlike Warner's design, the following design was used that was introduced by the American psychologist and Statistician Robert F. Boruch:

- With a predetermined and known probability p_3, the sensitive question "Did you cheat in the basic course?" should be answered;
- with a probability p_1 the respondent should simply give the answer 'yes,' and
- with the remaining probability $p_2 = 1 - p_1 - p_3$, the answer 'no.'

Thus, in contrast to Warner, the 'yes'- and 'no'-responses could be 'disguised' to different degrees by different choices of probabilities p_1 and p_2, meaning that the real 'yes'/'no'-answers are obscured by 'forced' 'yes'/'no'-answers with different probabilities p_1 and p_2. One possible random mechanism for selecting which instruction to follow without the usage of a randomization device, which can be applied to all survey methods, whether face-to-face, via mail, telephone, or even online, is based, for example, on birth dates:

Think of a person (yourself, mother, friend, ...) of whom you know the birth date. Now read the following instructions.

Instruction 1: If the selected birth date is in January or February, then just answer 'yes' at the end.

Instruction 2: If the selected birth date is in March or April, then simply answer 'no' at the end.

Instruction 3: If the selected birth date is from May to December, then truthfully answer 'yes' or 'no' at the end to the following **question**:

"Do you belong to the group of those who have violated the rules of the exam?"

Follow these instructions and give your answer now!

For simplicity, we assume that the probabilities are the same for all days of the year and we also neglect a possible leap day. Then the probabilities p_1, p_2, and p_3 for these three instructions are the following (rounded to two decimal places):

$$p_1 \approx 59 : 365 \approx 0.16$$
$$p_2 \approx 61 : 395 \approx 0.17$$
$$p_3 \approx 245 : 395 \approx 0.67$$

According to this procedure, a 'yes'- or 'no'-answer may indicate that the respondent thought of a birthday from May to December and actually gave the answer to the sensitive question, but it may as well mean that a birth date was thought of that was in the first two months of the year (or March or April, respectively) and so the respondent had to answer 'yes' (or 'no') only on the basis of following instruction 1 (or 2, respectively). This 'disguises' both possible answers to the sensitive question to a certain degree, depending on the probabilities chosen, as motivated at the beginning of this section.

With respect to the estimation of the proportion of students who knowingly violated the exam rules, the indirect questioning revealed the following information:

- 20 of the 64 participants in the course ticked the answer 'yes' on their questionnaire,
- 44 the answer 'no.'

What can be concluded from this about the behavior of all 64 students? Based on the given probabilities, it was to be expected that of all 64 people (Fig. 22.1)

- 16%, i.e. $N \cdot p_1 \approx 64 \cdot 0.16 \approx 10$ persons, had to follow instruction 1 (='yes');
- 17%, i.e. $64 \cdot 0.17 \approx 11$ persons, had to follow instruction 2 (='no'); and
- 67%, thus $64 \cdot 0.67 \approx 43$ persons, had to follow instruction 3 and consequently answer the sensitive question.

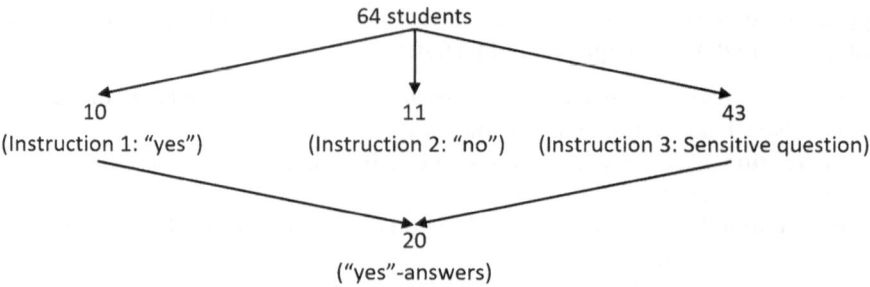

Fig. 22.1 Estimation of the proportion of individuals who violated the exam rules with an indirect questioning design

In Fig. 22.1, let us take a closer look at the randomly selected 43 persons who had to answer the embarrassing question: how many of them answered 'yes?' Out of the total of 20 'yes'-answers, ten of them were the result of following instruction 1. The remaining ten of the 'yes'-answers must have been given by the expected 43 persons who had to answer the sensitive question. Thus, we estimate the proportion of individuals who violated the exam rules in this group of 43 students by

$$10 : 43 \approx 0.23 = 23\%.$$

22.3   Tasks for Future Research

For the practical application of these theoretical considerations two prerequisites should be met: acceptable interviewer and respondent effort (without wheels of fortune and other devices, for example) and an easily applicable unified theory underlying the indirect questioning designs. To guarantee the latter, standardization procedures that combine such designs under a single theoretical umbrella can help. For example, the various instructions from Warner and the technique applied in the experiment described in Sect. 22.2 can be combined into one list of instructions. If only a subset of all instructions is used, for example, the three from our experiment, the probabilities of the unused instructions (in this case the opposite of the sensitive question in Warner's design) are simply set to zero. For example, if one wishes to supplement Warner's technique with instruction 1 from our experiment in order to protect the 'yes'-response more strongly than the 'no'-response, the instruction to simply say 'no' would be given a probability of zero. Therefore, there is no need to utilize and describe a new theory for each individual questioning technique that belongs to such a family of methods. In addition, such a unified theory should also allow for all probability sampling methods, from simple random sampling to probability-proportional-to-size random sampling.

Another interesting research question concerns the degree of objective privacy protection as a function of the design probabilities. Clearly, only if the level of privacy protection is the same, different indirect questioning techniques should be compared in terms of the accuracy of the survey results. The objective protection by the questioning design can be contrasted with the respondents' subjectively perceived privacy protection. Whereas the former directly affects estimation accuracy, it is the latter that has been empirically shown to control the respondents' willingness to provide reliable answers.

22.4 Further Reading

The basic idea of this chapter can be found in Stanley L. Warner (1965): "Randomized response: a survey technique for eliminating evasive answer bias," Journal of the American Statistical Association 60, 63–69. Unification of the methods described above can be found in Quatember, A. (2016): "A Mixture of True and Randomized Responses in the Estimation of the Number of People Having a Certain Attribute," in Chaudhuri, A., T. C. Christofides (eds.), "Handbook of Statistics (Volume 34) – Data Gathering, Analysis and Protection of Privacy through Randomized Response Techniques: Qualitative and Quantitative Human Traits," Amsterdam: Elsevier, 91–103 for all types of variables such as categorical (for example, party preference) or metric (for example, extent of tax fraud). On privacy protection, see Quatember, A. (2019): "A discussion of the two different aspects of privacy protection in indirect questioning designs," Quality & Quantity 53(1), 269–282, DOI: 10.1007/s11135-018-0751-4. For a practical application of indirect questioning, see, e.g., Krumpal, I. (2012): "Estimating the Prevalence of Xenophobia and Anti-Semitism in Germany: A Comparison of Randomized Response and Direct Questioning," Social Science Research 41(6), 1387–1403. On the effect of indirect questioning on the quality of survey results, see Wolter, F. P. Preisendörfer (2013): "Asking Sensitive Questions: An Evaluation of the Randomized Response Technique Versus Direct Questioning Using Individual Validation Data," Sociological Methods & Research 42(3), 321–353.

Chapter 23
Samples and Missing Data

Andreas Quatember

Abstract Textbooks on sampling theory assume unambiguous and easy-to-describe variables and survey respondents willing to provide the required information. Since in many disciplines this often does not apply, certain 'repair procedures' are necessary. Modern statistics shows how these work.

23.1 Sampling in Theory and Practice

"Only one in three people is satisfied with the work of the federal government", "poverty is at its highest level within ten years time", "60 percent of the population feel confident about the coming year"—headlines like these are usually based on results from sample surveys. Concluding from a sample to the wider population requires making so-called inferences. In order for such inferences to succeed, it is usually left to chance which individuals are included in the sample. Classical sampling theory provides the theoretical background for such inferences based on probability theory. However, theories often only apply under ideal conditions. In sampling theory, this means that a list of the studied population must exist from which a sample can be drawn. Moreover, each sampled person would have to be available and both willing and able to participate.

In practice, however, this is rarely the case. Almost all surveys suffer from high rates of non-response and, hence, missing data. Therefore, methods had to be developed that take into account realistic survey conditions. In practice, the actually drawn sample S is divided into three different sets with respect to the study variable (see Fig. 23.1):

- the part R of all those who responded (the 'response set');

A. Quatember (✉)
JKU Linz, Institute of Applied Statistics, Linz, Austria
e-mail: andreas.quatember@jku.at

© The Author(s), under exclusive license to Springer-Verlag GmbH, DE, part of Springer Nature 2024
C. Weihs et al. (eds.), *Statistics Today*, Society, Environment and Statistics, https://doi.org/10.1007/978-3-662-68907-3_23

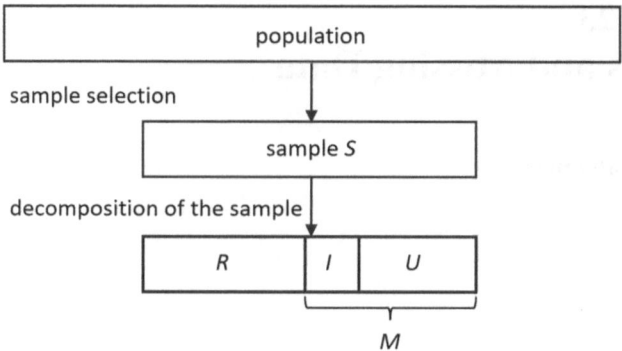

Fig. 23.1 Division of a sample S in the presence of non-response into a response set (R) and a missing set (M), which consists of the item non-response set (I) and the unit non-response set (U)

- the part I of those who participated in the survey but did not give an answer concerning the particular characteristic (the 'item non-response set'); and
- the part U of those who had been chosen for the sample but could not be reached or who had refused to participate in the survey (the 'unit non-response set').

The parts I and U are summarized as the 'missing set' M.

23.2 Statistical Methods to Compensate for Non-responses

As an example, we consider the estimation of the total value of a variable in the whole population (e.g., the monthly consumption expenditure of all households or the number of unemployed persons in a country) from a probability sample. Since not all persons of the population are included in the sample, the total of interest has to be extrapolated from the sample values. This is realized by a sum of weighted sample values of the variable. This means that, in the sum, the sample values are multiplied by a weight. This 'design weight' is larger than 1 and, as a rule, equals the reciprocal value of the probability for respondents to be included in the sample. For example, if the probability of a person to be chosen is $1/1000$, then the design weight is 1000. In this way, a chosen person would represent the responses of 1000 people of the population. However, under the previously described real conditions, there is the problem that the sum of the weighted data cannot be calculated from the entire sample, since a part of the needed sample values, namely those values in the missing set M, was not observed. The sum of values is actually only available for the response set R. This problem might also occur in complete population surveys. Therefore, as early as in the design phase of a survey, everything should be done to ensure that both the item non-response rate and the unit non-response rate are as low as possible, because no statistical method to compensate for non-responses can compete with a survey with a complete set of responses.

One methodological tool to compensate for non-responses is the 'subsampling' approach. The idea is to draw a random subsample from the missing set M of the original sample S with a response rate as high as possible. The results of this subsample are extrapolated to the complete missing set and from there to the total population.

However, if the non-response rate cannot be further reduced and, due to the number of non-responses, not be neglected either, two further methods can be employed to deal with the problem. As the basis for these techniques, auxiliary information is utilized that would otherwise remain unused. We will describe different kinds of auxiliary information in the next paragraphs.

The first method to compensate for non-negligible non-response was used in the field of official statistics as early as in the 1940s. It uses the values of the variable of interest observed in the response set and tries to estimate the total value in the whole population based on this set alone. For this purpose, it is necessary to increase the originally intended weights accordingly. This 'weighting adjustment' will be successful if it is based on a plausible assumption about the mechanism that generated the non-response. In practice, a distinction is made between three possible non-response mechanisms. These mechanisms represent one kind of utilization of auxiliary information, the other kind will be discussed in the next paragraph. The first mechanism considers the non-response to be completely random. Under this assumption, the response set is a random sample from the original sample. In this case, the data of the response set are weighted uniformly higher simply because of the reduced sample size. In the second mechanism, it is assumed that non-responses depend on the values of variables available in the original sample for all selected individuals. For example, if non-responses depend on the variable 'gender', i.e. men have different non-response rates than women, men would receive different weights. The third possible non-response mechanism is by far the most unpleasant one for the researcher in the process of compensation of non-responses. In this case, non-responses do not depend on observable variables alone, but also on the value of the variable of interest itself. This is the case, for instance, when in a household survey on consumption expenditure non-responses can be explained not only by the number of household members (the more of them, the more likely someone will provide the requested information), but also by the households' consumption expenditure itself (especially households with high expenditure may refuse to provide information). In such cases, the assignment of different weights according to household sizes might not help to compensate the bias in the observed sample.

The other possible method to compensate for non-negligible non-responses besides the method of weighting adjustment estimates the missing responses of individuals in the missing set on the basis of their similarity (with respect to some characteristics such as gender and age) to persons in the response set. Based on the observed data from the response set, the missing data are replaced by estimates, which are used to extrapolate the total sample to the population. Of course, the quality of such 'data imputation' depends on how accurately the mechanism used in the imputation process reflects the mechanism that really caused the non-response. As for the method of weighting adjustment we will discuss three kinds of mechanisms used in imputation processes. One possible imputation method is the random assignment of observed values from a similar group of responding individuals (for example, of the same gender and age) to the missing set. Here, the values in the similar group are used as auxiliary information. This method is called 'hot deck imputation' because it was first used at a time when data were stored on punch cards. The needed data for the imputation process were thus taken from a stack of punched cards from the same deck, which was still hot. A 'cold deck imputation', accordingly, uses information on missing values from other sources such as previous surveys (consumption expenditure of the same household in the previous month, for example). In this method, the values from the other sources are the auxiliary information. Another obvious possibility is to use the data of the response set to estimate a regression equation based on the observations of variables available for all sample persons (e.g. age, educational level, etc.) and to impute the missing values of the persons in the missing set based on this regression. If the statistical model is correct, the imputed values can be used to obtain estimates of, for example, totals that are less biased than estimates based purely on the data available in the response set. In this method, the observations available for all sample persons are used as auxiliary information.

Weighting adjustment and data imputation are not in competition. The latter method is more suitable for compensating for item non-response which occurs when a person participating in the survey refuses to provide information on one or more, but not on all survey items. Therefore, auxiliary information about other items is available in this case. The former method, i.e. weighting adjustment, however, is particularly suitable for compensating for unit non-response when, as a result of the participant's non-response, no additional information other than that officially available can be obtained from the sample person. Since both types of non-response occur in most surveys, the two methods are usually carried out in succession, so that compensation is first made for item non-response by data imputation. Then it is also carried out for unit non-response by weighting adjustment on the basis of the response set and the item non-response set filled by the imputations (R combined with I). This is the approach taken, for example, in the large sample surveys of official statistics, carried out by statistical offices, such as the Microcensus and the EU Labor Force Survey.

23.3 Further Reading

The 'subsampling'-method goes back to M. H. Hansen and W. N. Hurwitz (1946): "The Problem of Non-response in Sample Surveys," Journal of the American Statistical Association 41, 517–529. The standard work on non-response is: R. J. A. Little and D. B. Rubin (2011): "Statistical Analysis with Missing Data," 2nd edition, New York, Wiley.

Part VI
Language Data

Chapter 24
Who Is Supposed to Read All This?
Automatic Analysis of Text Data

Jörg Rahnenführer and Carsten Jentsch

Abstract Statistics deals with data, and data consists of numbers. That is what most people associate with statistics. However, with the IT revolution, further, quite different things have come into the focus of statistics as a science, including texts. Even though these have been generated in large quantities from a wide variety of sources for many centuries, they have become widely available with the increasing digitization of texts only recently and can nowadays be accessed as parts of large text collections. This has made it possible to analyze the information contained in text data in a structured and automated way. To this end, statistics provides many helpful methods, both for processing the data and for interpreting it.

24.1 Large Text Collections

In recent years, large text collections have become available in digital form in a variety of different forms. Examples include newspaper articles from a particular daily or weekly newspaper, online text collections, and even online encyclopedias. Almost any other source of text data can also be of interest for content analysis, such as election programs of political parties or written speeches in parliament.

Nowadays, we might be confronted with the task of approaching and analyzing large text collections, for example as part of our studies or jobs. However, if we want to gain general insights from these large volumes of text that go beyond the individual text, it is usually not possible to read all texts and draw unbiased conclusions from them in a subjective way. Such insights are desirable, though, for example, in order to analyze the overall coverage of a particular topic: Who participates in public discourse on what topics and when?

Looking at the election programs of parties that have existed for decades, for example, the question may be of interest, how parties can be placed in the political

J. Rahnenführer (✉) · C. Jentsch
TU Dortmund, Department of Statistics, Dortmund, Germany
e-mail: rahnenfuhrer@statistik.tu-dortmund.de; jentsch@statistik.tu-dortmund.de

© The Author(s), under exclusive license to Springer-Verlag GmbH, DE, part of Springer Nature 2024
C. Weihs et al. (eds.), *Statistics Today*, Society, Environment and Statistics, https://doi.org/10.1007/978-3-662-68907-3_24

spectrum over time and which topics determined the political debates and when. In recent decades, the possibilities of algorithm-based analyses of text data have increased enormously, and automatic evaluations of these data are now common in many fields. However, these analyses are subject to all kinds of uncertainty, and this is where statistical methods can help to evaluate such data in a more meaningful way.

24.2 Text Analysis in the Social Sciences

In terms of content, the social sciences deal with questions of how people live together in society. In this context, texts play a major role in different types of media, for example in books, journal or online articles, in social media, or in interviews that are available as texts. One method used in the social sciences to understand texts is content analysis. With this, texts are evaluated quantitatively. This is usually done by reading and evaluating specific texts, or at least a random selection of them, according to predetermined rules. This produces structures in the data that, for example, helps to classify those texts into categories, or to count the frequencies of, for example, certain words or grammatical constructions.

It is important to ensure that researchers working with the texts and assigning them to categories apply objective and comparable criteria. To check this, statistical measures can be used to verify that researchers come to reliable and comparable results. The resulting data set can then be used to investigate questions from the social sciences. For example, statistics can be used to investigate whether the proportion of certain grammatical constructions differs in different subject areas (politics, economics, sports, etc.).

In the past, the selection of data sets for such analyses was often limited to ensure that all selected texts could be read and categorized by the researchers. For example, only a single week of coverage in a particular newspaper was considered. With computer-assisted text mining, it is now possible to evaluate even very large collections of text.

In addition to classical content analysis, many other types of text analysis in the social sciences benefit from at least partially automated analyses. The following two examples are related to political elections. As a first example, online comments or Twitter messages (tweets) about the top candidates in an election should be compared in terms of approval or disapproval of the candidates. As a second example, the positioning of parties in the political spectrum as well as their changes over time should be determined based on the identification and counting of important topics in their party programs. Both, the evaluation of all comments and the counting of specific words in the party programs, cannot be done manually, i.e. without automated procedures. See Sect. 24.6 for a more detailed discussion.

24.3   Preprocessing of Text Data

Figure 24.1 shows a pile of newspapers that a human being cannot read completely and for which a full quantitative analysis would not be possible. Therefore, the articles have to be made machine readable.

Articles from newspapers are available in large archives as text data in a structured format that contains, in addition to the article text, information such as the publication date, the title, or the author. After they have been read in, however, the computer is not yet able to handle the texts well. While a human can easily extract the important information from a text, texts must first be tidied up and structured for a computer. First, they must be checked for peculiarities, for example, whether some texts are implausibly short or long, or occur more than once in the data set. The former is done using statistical methods, for example histograms that show the length distribution across all texts or statistical tests that identify texts of extreme length as so-called outliers.

Often, for further automated processing, the bag-of-words approach is considered, in which only the frequency of words is counted for each text but the order of words is neglected in order to be able to better process and compare the texts. In addition, there are also approaches such as part-of-speech tagging or entity recognition, which attempt to recognize the sentence structure (subject, verb, object) or the entities described in a text (proper names, company names, cities, etc.). Statistical models are used here as well.

A typical preprocessing step in the bag-of-words approach is the removal of so-called stop words that are assumed to have no meaning in terms of content by themselves, but are important for the reading flow, e.g. articles (*the, a, an*) and frequently used conjunctions (*and, or, when*). Furthermore, punctuation marks and numbers are often removed and all upper case letters are converted to lower case. Finally, all spaces are considered as separators between words to obtain a word list. This step is called 'tokenization'. After these simplifications, the content issues can be addressed.

Fig. 24.1 Pile of newspapers: Who is supposed to read all this and identify the essential statements? (see https://pixabay.com/de/photos/bundle-jute-seil-zeitung-1853667/ [Pixabay licence])

24.4   Topic-Based Classification of Large Text Collections

As already mentioned in the content analysis in Sect. 24.3, the topic-based classification of individual texts into content-related subgroups is an important goal when processing large text collections. For example, one is interested in all texts of a daily newspaper that deal with the topic *bank*. Here, the bank is meant as a financial institution, which typically manages current accounts or grants loans. A simple and popular solution is to select all texts in which the word *bank* appears at least once. However, this often does not lead to the desired result, as both unwanted articles are found (false positives) and articles are overlooked (false negatives). The first case occurs when *bank* is part of a *blood bank* or *data bank*, for example, or the word *bank* is part of another word, such as in the word *emBANKment* (a bank or wall made of stone or earth supporting a roadway or holding back water). The second problem can occur when instead of the word *bank*, for example, the word *financial institution* is used in the article.

A specific statistical method can take into account both problems and partially solve them, the so-called LDA method (Latent Dirichlet Allocation). For this purpose, a simplified model is assumed that works with probabilities and that models texts as mixtures of different content topics. For example, if an article deals with a *bank* as a financial institution, then words such as *credit, money*, or *stock exchange* occur more frequently than in a randomly selected article. If it is about *embankment*, then words like *sand* and *water* occur more frequently.

The exact probabilities that words are selected for a certain topic are estimated from large text collections using the LDA method. Texts can then be assigned to topics. The fact that topics are actually estimated automatically without prior knowledge seems very surprising at first glance. This is because texts usually consist of a few topics that can be identified with the help of large text collections, since texts on the same topics normally use quite a number of similar words.

However, there are other statistical challenges in categorizing texts into topics. In particular, the number of topics in texts is normally unknown. When getting started, the LDA method nevertheless requires a value for the number of topics into which texts should be divided and for which a good solution should be found. However, on the one hand, the most adequate number of topics depends on statistical factors, such as how similar two texts on the same topic must be at least. On the other hand, it depends on content-related questions such as whether articles on a certain, rather special event should be assigned to a separate subgroup.

Once one has extracted the topics and the assignment of texts to them from the data, one can use this to tackle many exciting questions. Example questions for text collections of newspaper articles or online articles are: Which topics are particularly popular in the text collection, what is reported or written about in the first place? How important is a particular topic over time? When did a topic first appear or reappear? For the last question, one can then check which events might have triggered this increased media coverage. The LDA method thus provides assignments of topics that can be examined by content analyses.

There are very many ways to classify texts according to topics and a large number of probabilities of texts belonging to the topics have to be estimated simultaneously for an LDA model. Moreover, in addition to the number of topics other numerical values must be specified for estimating topic classification, for example how many topics occur on average in a single text or what the average number of important words is for a topic. This leads to yet another reason why statistics is needed, namely to check the stability of the complex model corresponding to the choice of all the auxiliary parameters. To this end, statistical methods can be used to test how much the result depends on the choice of these parameters, i.e., how topics may change when other parameter values are chosen. Checking the dependence of the model on the choice of such parameters is a core competence of statistics and shows how much one can trust the very complex estimated model.

24.5   Finding Differences

Many statistical analyses aim to find differences between two groups. For example, if the reduction in blood pressure is measured for two groups of patients with different medication, or if the number of defaulted loans is measured for two different groups of banks, then two sets of numbers are available in each case. For these, in the simplest case, only mean values (of the reduction in blood pressure or of the number of loan defaults) and estimated variations of these values in the groups need to be calculated. Then, it can be determined whether the differences are stronger than would be expected from purely random fluctuations.

For the text data described here, this is much more complicated. The result of an LDA analysis are topics described by probabilities of important words related to the topics. If one wants to compare topics in one newspaper with topics in another newspaper, one must first match the topics found in the two newspapers and then analyze the differences in the frequencies of texts related to the topics in the two newspapers. However, in this respect, a number of statistical questions and challenges exist.

1. How do we calculate the distance between two topics?
   Topics are described by so-called probability distributions, i.e. by probabilities assigned to words. To calculate the distance between two topics, one has to calculate the distance between the corresponding probability distributions. To this end, the probabilities must be compared across all words, for which there are tailored statistical distance measures.
2. How do we measure the statistical relevance of the calculated difference?
   The algorithm for measuring the statistical relevance of the differences calculated in the preceding step is very complex, and there is no simple direct way (as in the examples at the beginning of this section) to measure whether the observed difference could also occur by chance. One technique from statistics that can help here is called bootstrap analysis. The term 'bootstrapping' relates to the

idea (and was labeled accordingly) of "pulling oneself out of the swamp by pulling ones bootstraps". For our example, this means that new, similar texts are created from the original texts. To this end, one randomly and repeatedly 'pulls' words or entire sentences from each individual text and combines them into a new text. This way, similar texts are created, which show random variations of the original texts. Then the original LDA analysis is repeated on the new texts. By considering the variation of the LDA models on the repetitions, one can estimate the statistical uncertainty of the LDA result. This is then used to evaluate the statistical relevance of the results, here of the differences between text collections.

24.6   Text Analysis of Election Programs

Political institutions regularly generate large amounts of textual data. This can be, for example, the written speeches of individual politicians in parliament, legislative proposals submitted to parliament, or election programs as published by all major parties for federal or state elections. Election programs, sometimes also called 'party manifestos', are of particular interest to characterize political change because they are written at regular intervals and address the parties' positions on all politically relevant issues. Thus, text collections containing election programs of several parties over many election periods represent a promising data source for automated text analysis.

The parties represented in the German parliament in 2021 were CDU/CSU (Christian Democratic Union of Germany and Christian Social Union in Bavaria; political position: center-right), SPD (Social Democratic Party of Germany; political position: center-left), die Grünen (The Greens; center-left), FDP (Free Democratic Party; center to center-right), Die Linke (The Left; former PDS, left-wing to far-left), and AfD (Alternative for Germany; right-wing to far-right). The AfD was only founded in 2013 and is thus not considered in the analysis below, since we analyze the political development from 1990 to 2013. As can be seen in Table 24.1, the election programs of many German parties have become increasingly long and complex over the years. While the length of the FDP's election programs has fluctuated only slightly over time, the length of the other parties' election programs has increased considerably. The Green Party's election program for the 2013 federal election, for example, is around 20 times longer than the one for the 1990 federal election. German print and TV media noticed this development in the run-up to the 2017 federal election (translations below):

- "Thick tons of election programs: Who's going to read it?" (ZDF, heute)
- "Reading all the election programs? Takes only 17 hours." (Die Zeit)
- "Election programs as thick as telephone books. Who on earth is supposed to read all that?" (Bild Zeitung)

Table 24.1 Total number of words used in the election programs of the five most important German parties in the federal elections from 1990 to 2013

	1990	1994	1998	2002	2005	2009	2013
CDU/CSU	3061	6083	4455	10,559	5813	14,243	22,518
SPD	3991	7634	7274	10,919	6718	14,369	22,459
Die Grünen	2235	16,169	2215	11,883	15,053	18,646	44,192
FDP	14,476	21,567	12,895	16,788	11,496	15,684	20,195
PDS/Die Linke	4935	3629	8113	7387	4035	10,701	20,029

Due to the large amount of text, it is desirable—if not inevitable—to analyze these data automatically. The goal of such an analysis would be to extract relevant information contained in the election programs and to draw conclusions about the parties as such, as well as the German political system. In this context, possible questions might concern shifts in the political system: Has the relative position of parties in the political spectrum changed over time? Is it possible to identify a shift to the right or to the left for a particular party? Other relevant questions concern political discourse: Which political issues were important at which election? Do differences between progressive and conservative parties exist in terms of their choice of words or topics addressed?

In principle, statistical methods such as LDA can be used on text data of any origin to extract information contained in the data. In addition, however, methods have been developed that are particularly suitable for analyzing political text data. The so-called Wordfish algorithm uses a bag-of-words approach and proposes a statistical model to locate parties in the political spectrum and analyze shifts over time. In doing so, this approach is completely automatic and is based solely on word counts, without having to identify potentially important words in advance. In fact, independent of that, the algorithm can identify informative words that, for example, help to distinguish progressive from conservative parties and thus to place them on the political spectrum.

Current research in the field of political text data analysis is concerned, among other things, with further developments of the Wordfish approach that cannot only map shifts in party-political orientations over time but also allows for political connotations of particular words to change over time. For example, there may be topics or political issues that are highly controversial between parties at a certain point in time but are resolved soon after. For example, issues directly related to the reunification of Germany and the associated upheaval in the political system were very important in the early 1990s. However, their importance declined significantly in the following years. This was expressed, for example, in a less frequent and less politically charged use of the word *federal state* and the abbreviations *FRG* (Federal Republic of Germany) and *GDR* (German Democratic Republic). Likewise, new topics that had not been on the political agenda before, such as the Euro crisis from 2010 onwards, continue to emerge.

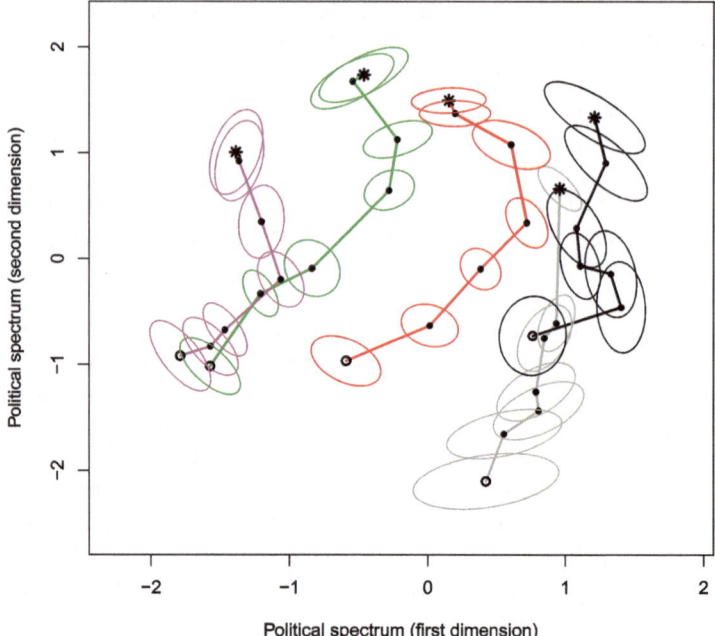

Fig. 24.2 Party positions of German parties in a two-dimensional political spectrum, CDU/CSU (black), SPD (red), Die Grünen (green), FDP (gray), and PDS/Die Linke (purple), in the years of federal elections from 1990 (°) to 2013 (*). Dots and asterisks represent mean party positions and ellipses represent statistical uncertainties

Other analyses allow, for example, for multidimensional party positions. Figure 24.2 shows the positions of the five major parties CDU/CSU, SPD, Die Grünen, FDP, and PDS/Die Linke in the programs of the federal elections from 1990 to 2013 in a two-dimensional graph. The closer two points are, the more similar the respective party programs are. It can be seen, for example, that the programs of the parties Die Grünen and PDS/Die Linke started to differ more clearly from each other with the Greens' participation in government from the end of the 1990s onwards.

Dealing with uncertainties in the application of such statistical methods is not easy, and the question of statistical relevance of, say, shifts in party positions in the political spectrum is generally difficult to answer. For this purpose, bootstrap procedures are usually used (for a description of these, see Sect. 24.5) to quantify uncertainties of the estimated modes. In Fig. 24.2, this can be seen from the ellipses around the points, which express the uncertainty in determining party positions in the two-dimensional representation.

24.7   Summary and Outlook

The automatic analysis of text data offers many new possibilities for an objective gain of knowledge. Many large collections of texts are increasingly generated by different media and are available for analysis. This is especially true for the still emerging online domain. While many methods have been developed for text mining, the field is still in its infancy. How to choose optimal parameters for algorithms and statistical methods is a subject of current research. Another promising idea is to include additional information in the analyses. For example, when aiming to identify topics, some texts may have already been assigned topics manually, which is a good starting point for an automatic analysis. It is also possible to use previous topic classifications from other text collections as a basis. How best to proceed in each case is an exciting challenge for the future.

24.8   Further Reading

In the field of text mining, a large number of publications and methods exists. A good overview is provided by Miner et al. (2012): "Practical Text Mining and Statistical Analysis for Non-Structured Text Data Applications," Elsevier Science & Technology. The LDA method described in Sect. 24.4 dates back to 2003. It was presented in the paper "Latent Dirichlet Allocation" in the Journal of Machine Learning Research 3, 993–1022, by Blei et al. The Wordfish algorithm for text analysis of election programs dates back to a paper by Slapin and Proksch (2008) in the American Journal of Political Science 52, 705–722.

Chapter 25
Statistical Modeling of Current Linguistic Realities Around the World: The Case of Singapore

Sarah Buschfeld and Claus Weihs

Abstract The English language has experienced an unprecedented spread in the last decades due to colonization and globalization. Nowadays, non-native speakers clearly outnumber native speakers of English, but a new trend towards new first language varieties of English has set in, in particular in Asian and African contexts. This not only changes the general character of the English language but questions old norms of correctness and standards for local and worldwide teaching contexts. By means of statistical modeling and on the basis of the example of Singapore, we show that English is not a monolithic whole (as often been assumed by non-linguists) but a complex language system, stratified by a number of social variables such as age, gender, and ethnicity.

25.1 Singapore and the World Englishes Paradigm

The study of 'World Englishes' is one of the striving sub-disciplines of modern linguistics—and rightly so. It aims to better understand our current worldwide lingua franca English, which influences the daily lives of the majority of the world's population. It is the first foreign language taught in most schools and the language of advertising, business, commerce, and tourism. Popular culture and new media are dominated by English. For an ever-increasing number of children, it is developing into (one of) their home language(s), even in contexts that formerly had languages other than English as their culturally and historically rooted mother tongue(s). Singapore, a modern city-state in Southeast Asia with a population of 6,209,660 (July 2020 est.), is the pioneer in this development. According to the 2020 Singapore

S. Buschfeld (✉)
TU Dortmund, Department of Cultural Studies, Dortmund, Germany
e-mail: sarah.buschfeld@tu-dortmund.de

C. Weihs
TU Dortmund, Department of Statistics, Dortmund, Germany
e-mail: claus.weihs@tu-dortmund.de

Census of Population, 77.4% of the Chinese, 63% of the Malay, and 69.8% of the Indian 5 to 14-year-olds nowadays speak English as the most frequently used home language (Department of Statistics Singapore 2020: 29).

Rising numbers and tendencies of convergence between the groups in recent years suggest a stabilizing trend and penetration of English in all segments of the population. Still, the different groups show important differences in their linguistic behavior, which is due to the fact that English has started out as a second language in Singapore. In such scenarios, the English language is always influenced by other languages spoken in the society.

In this chapter, we show how statistical analyses can help unveil the heterogeneous character of English and that even in a single society we find different varieties of the same language, mainly influenced by social variables such as ethnicity, age, and gender. We finally discuss what the findings imply for our understanding of language in general and the English language in particular and what repercussions this has for traditional linguistic standards and models for language teaching.

25.2 Data Collection and Preparation

The data were collected in Singapore and England in 2014 and 2015. Data from Singapore came from 30 children, male and female, aged 2;5 (two years; five months) to 12;1. Twenty children are of Chinese, 9 of Indian, and 1 of 'mixed' ancestry. The Malay group was excluded from the study for practical reasons. Data from England came from 21 children aged 2;1 to 10;9. Thirteen of these children are monolingual from 'traditional' English families, i.e. both parents were born in England and are first language speakers of British English. Six children are of migrant background, i.e. their parents are both speakers of other languages and not English. Two children are of mixed background, with one parent being a native speaker of British English and the other parent a native speaker of a language other than English. The children of migrant and mixed background are all either bi- or multilingual. The data were elicited systematically in video-recorded task-directed dialogue between researcher and child, consisting of several parts: a grammar elicitation task, a story retelling task, elicited narratives, and free interaction. The recorded material was orthographically transcribed and manually coded for the two grammatical features under investigation here:

1. the realization of subject pronouns (realized vs. zero) and
2. the use of past tense marking (marked vs. unmarked and verb+'finish' as past tense marker).

Realized subject pronouns and marked verb forms (both regular and irregular) which are to be found in the traditional native speaking varieties of English (British and American English) are illustrated in Examples 1 through 3 (taken from the Singapore data, but turned into their standard realizations):

1. Researcher: [. . .] what do you do with your friends? Do you play with them? Child: I play with them. [. . .] Sometimes **we** play some fun things.
2. Child: Then he **wanted** to climb a ladder to a chimney. Then the big bad wolf **was** in the pot. Then all the water **splashed** and the carrot and the onion.
3. Child: He **ate** everything.

Examples 4 through 6 illustrate the respective Singapore English (SingE) variants, viz. zero subjects and unmarked verb forms as well as a very specific SingE variant (verb-finish). They are what the children actually uttered in these contexts.

4. Researcher: [. . .] what do you do with your friends? Do you play with them? Child: [Ø **I**] Play with them. Sometimes drawing. Child: Sometimes [Ø **We**] play some fun things.
5. Child: Then he wanted to climb a ladder to a chimney. Then the big bad wolf **is** in the pot. Then all the water **splash** and the carrot and the onion.
6. Child: He **eat finish** everything.

The forms in Examples 1 through 6 are variably used by children acquiring English as a first language in Singapore. The SingE variants illustrated in Examples 4 through 6 are also typical for adult SingE and are thus part of the input the children receive. These features are the product of general mechanisms of language acquisition such as simplification and L1 (first language) transfer from the local languages of Singapore, viz. Chinese and Indian languages and Malay. This means that second-language learners—and this is what the Singaporean population originally was—take over linguistic characteristics of their L1 to their L2 (here English) even if they do not match the target grammar. This is how new varieties of English are born.

The aim of the analyses is to find prediction rules for the use of subject pronouns (realized vs. zero) and for past tense marking (standard vs. nonstandard subsuming unmarked and finish, viz. the SingE variants) by means of extra- and intralinguistic predictors.

The extralinguistic features considered as independent variables in the statistical analysis are ETHNICITY (ETH), AGE (AGE), SEX (SEX), LINGUISTIC BACKGROUND (LiBa), and MEAN LENGTH OF UTTERANCE (MLU). MLU is an aggregated factor for which children were assigned to three groups according to the average grammatical complexity of 50 of their utterances. The intralinguistic feature modeled in the subject study is PRONOUN (PRN), and we looked into whether the exact type of pronoun (*I, you, he, she, it, we, you, they*) has a statistically significant influence on the results. For the past tense study, we considered VERB TYPE (VT: regular vs. irregular) as an intralinguistic predictor.

25.3   Prediction of Linguistic Characteristics

The type of problem we consider here is a classification problem, since the target to be predicted is a class variable with two levels each in both problems. The target variable is then modeled to be predicted by the extra- and intralinguistic features. The results are classification rules, which allow for generalizations to children with similar backgrounds growing up in England and Singapore.

To analyze the impact of the predictors, we modeled conditional inference trees (ctree), using the package 'party' in the software R, in order to predict outcome classes. A ctree analysis investigates a data set in a recursive fashion to determine an optimal series of significant splits to predict the realizations of the dependent variable. A (binary) split divides the data into two groups with regards to the realizations of a (categorical or numeric) independent variable. Ctrees are currently considered one of the best-practice statistical methods for linguistic analyses. They are widely used in the field for a variety of reasons (often in combination with mixed-effects models and random forests): the final result is a readily accessible and interpretable decision tree, which straightforwardly visualizes the interplay of multiple predictors. As another advantage, ctrees are not overly sensitive to outliers, which is a common phenomenon found in linguistic studies.

25.3.1   Design of Experiments

The studied classification problems are heavily unbalanced. In the subject study, 528 tokens were categorized as 'zero' and 5618 tokens as 'realized', in the past tense study 640 tokens were categorized as 'nonstandard' and 1549 tokens as 'standard'. This means that the size of the smaller class in the subject study is only 8.6% of the overall data set; in the past tense study it is 29.2%. In cases with an extremely small smaller class, the 'baseline classification rule' in the two-class case is "always take the larger class", which would lead to an accuracy rate of around 90% in our subject study. With this rule, not a single token of the smaller class has been classified correctly. In order to find classification rules able to also adequately classify the smaller class, an evaluation measure has to be used which equally takes into account the accuracies in all classes. This can be realized by means of the so-called 'balanced accuracy', that is the mean of the accuracies of all classes. In what follows, we will rely on this measure.

In order to construct a classification rule with acceptable balanced accuracy, we apply the method of 'undersampling'. Undersampling takes the full sample of the smaller class together with a small sample of the larger classes as the 'training sample', from which the chosen classification method constructs the rule.

The 'overall balanced accuracy rate' is then computed on all observations of both classes. For the smaller class, the accuracy is computed on the training sample, i.e. the full data sample of the small class. For the larger class, it is computed not only on the training sample, but also on the 'hold out' from rule construction. This way, rules from different training samples can be adequately compared and the best rule can be identified by looking for the highest balanced accuracy.

The sample size of the larger class used for training can be varied. We evaluated different sample sizes. We only show the results for the sample size with the best overall balanced accuracy. In the subject study, the best balanced accuracy was 70.2%, in the past tense study it was 76.0%. These values are reasonably high for problems with such differences in class size.

25.3.2   Variation in British and Singaporean Englishes

Past Tense Marking

We start by discussing the results from the past tense marking analysis. In order to improve the balanced accuracy of the model, for this analysis we excluded observations with MLU = 'OL' (outlier) and LiBa = 'NA' (not available). The best ctree model after 151 repetitions of undersampling, based on 50% of the large class, and a significance level of 1% is presented in Fig. 25.1.

Linguistic background (LiBa) appears to be the strongest predictor for the variation in the data. The most important split (node 1) divides the data into two major classes, viz. children that grow up with two languages from before age two (bili1 and multi1) and children that are either monolingual (mono and mono+) or have acquired additional languages after the age of two. Both classes show further splits according to LiBa further down in the tree and in interaction with other predictors. On the right side of the tree, those monolingual children 6 years and older that have started learning an additional language in school (mono+) split from the rest and show almost exclusively standard realizations of past tense verb forms. This is by no means surprising since the mono+ group contains English ancestral children (E/a) only, this means children growing up with both parents being native speakers of English. The next split (node 18) illustrates a statistically significant difference between male and female speakers, with the female speakers showing more standard speech behavior. Sociolinguistic research of the last 60 years has repeatedly shown that females often use more standard speech forms than men. This has been explained on the basis of a number of female attributes, such as their status in society, conservatism, prestige consciousness, upward mobility, insecurity, deference, sensitivity to others and to social implications of speech, etc.

In the female group, two age splits occur, with those children 84 months (7;0) and younger (node 20) and 75 months (6;3) and younger (node 23) showing higher frequencies of non-standard realizations than the older ones. This is a general feature of early stages of language acquisition, no matter what variety of English children acquire and whether they are mono- or multilingual (node 19).

The male group splits into monolingual speakers on the one hand and late bilingual speakers (LiBa = bili2) on the other, with the latter using non-standard

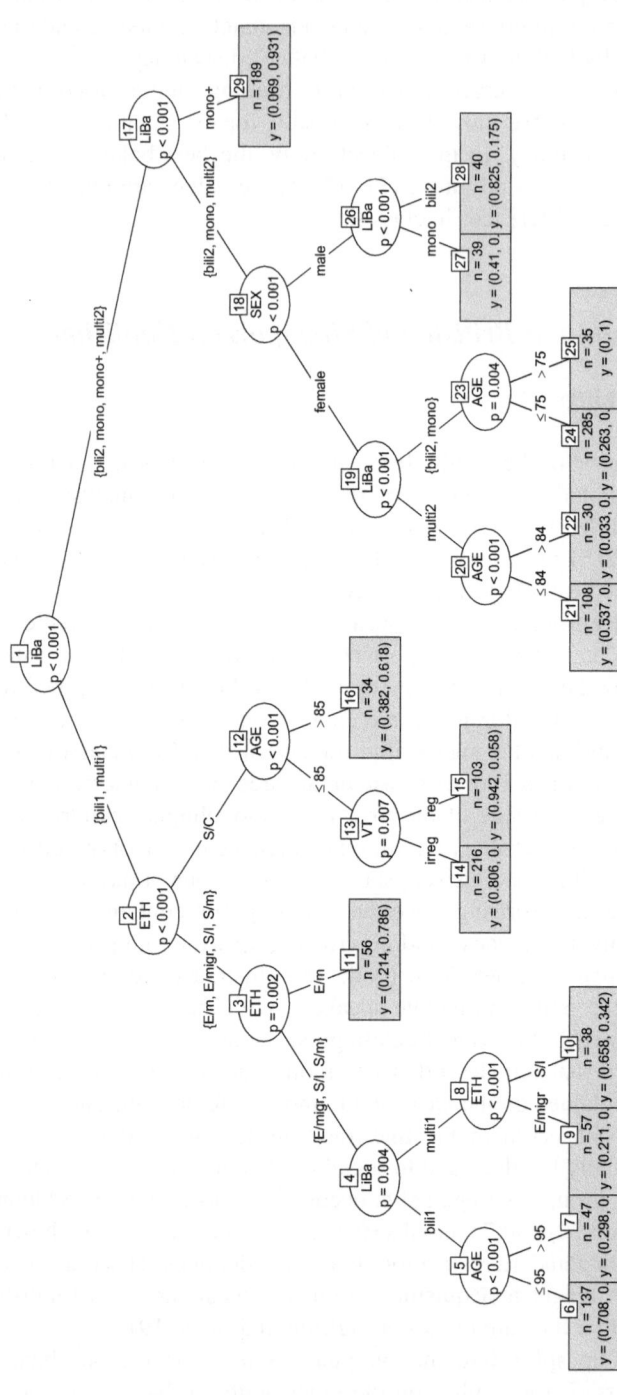

Fig. 25.1 Conditional inference tree for past tense marking; simple version of terminal nodes: y = (no.nonstandards, no.standards)

forms much more frequently. This is most likely the result of cross-linguistic influence, which is quite typical in bilingual language acquisition. In the present case, the child's other language is Chinese, which does not realize inflectional endings at all.

The left side of the tree splits the early bi-/multilinguals into the group of Singapore Chinese on the one hand and all other children on the other except for the ancestral children in England which are all monolingual and therefore did not make it into this part of the tree. For the Chinese Singaporeans, the data set is then split by AGE (node 12) into children older than 85 months (7;1) and those 85 months or younger. This corresponds to the age split identified for the right side of the tree in that, again, the older ones realize the non-standard forms less frequently than the others. The split is well-motivated since children in Singapore start compulsory education at around the age of 6. From then on, they are more strongly exposed to standard speech forms due to a rigid, Standard English oriented speech policy of the Singaporean government. However, node 16 shows that non-standard realizations still play a prominent role. The final split in this branch relates to VERB TYPE. Most of the children employ quite high frequencies of non-standard forms for both verb types.

The split in node 3 again divides the children according to their ethnicity, with the children from mixed families in England behaving more standard-like than the rest. This is most likely due to the fact that at least one parent is a native speaker of British English and constantly provides the standard input. The children from Singapore, together with the migrant group growing up in England are then split by linguistic background, with the multilingual Indian children from Singapore (node 10) and those bilingual ones 95 months (7;11) and younger producing exceptionally high numbers of non-standard past tense structures.

Subject Pronoun Realization

For this study, we aggregated all monolingual children (mono and mono+) to 'mono' and all others to 'multi' in LiBa. Again, we used 151 repetitions, this time with 9% undersampling of the larger class, and, again, a significance level of 1%. The resulting tree is shown in Fig. 25.2.

The subject pronoun analysis generally confirms the findings of the past tense study in that age, ethnicity, linguistic background, and sex have a similar impact on whether the children realize subject pronouns. Other than in the past tense tree, however, the intralinguistic predictor 'pronoun type' has the strongest impact on the realization of subject pronouns. Zero pronouns most prominently occur for the pronoun *it*, this means either the semantically empty dummy *it* ("It is raining") or the referential *it* ("My cat, it is black"). In both cases, the pronoun *it* is frequently deleted in structures such as *it's* due to phonological assimilation. Still, realization of zero subject pronouns depends on the speakers' ethnicity (node 13). This time, the Singapore Chinese and Indian children have far higher zero subject rates than all other children.

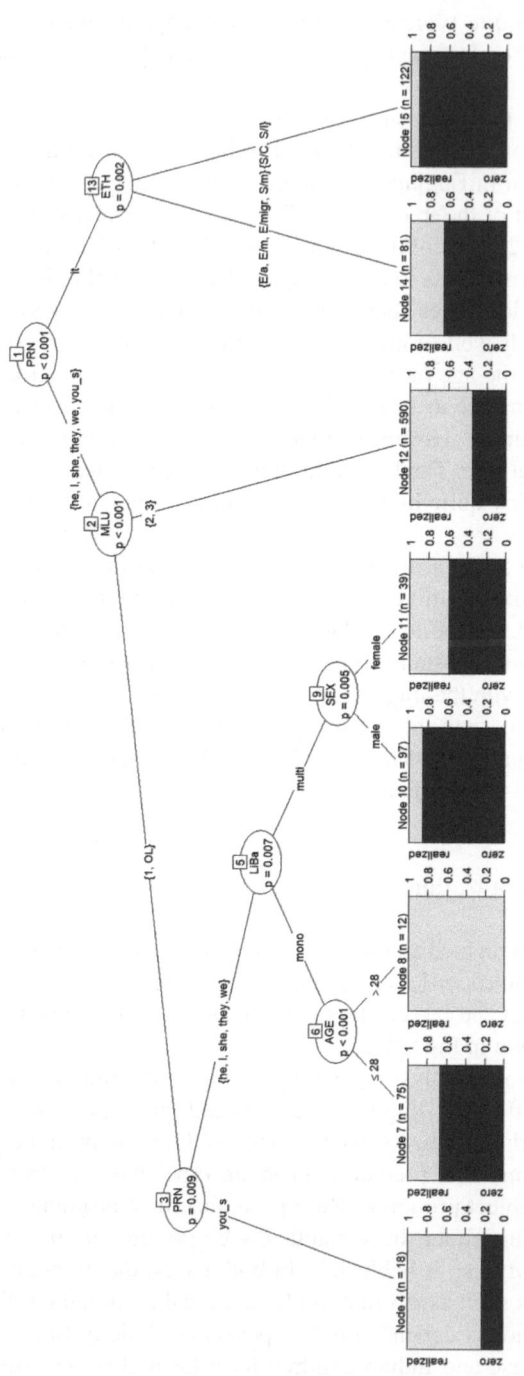

Fig. 25.2 Conditional inference tree for subject pronouns

For the remaining pronoun types, an age-related effect can be seen in split 2 in the left side of the tree (MLU relates to age). Again, the younger children use significantly higher numbers of zero subjects and thus non-standard forms. Node 3, again, splits the data according to pronoun type with singular *you* (*you_s*) being realized as zero less frequently than pronouns *I, he, she, they, we*. The realizations of the latter depend on further variables. Multilingual speakers use significantly more zero pronouns than monolingual speakers (node 5) and male speakers, again, more than female speakers (node 9). The monolingual speakers are again split by age (node 6).

Overall Findings

Overall, we found that age, ethnicity, linguistic background, and sex have a similar impact on both past tense marking and subject realization. In general, Singaporeans, and in particular those of Chinese descent, use non-standard speech forms more frequently than children growing up in England. Beyond that, bi- and multilingualism is an indicator for non-standard forms. Furthermore, male speakers show a stronger inclination towards non-standard realizations than females, and younger children than older children. Therefore, language use is clearly influenced by biological and sociolinguistic factors.

25.4   Evaluation and Interpretation

Finally, we discuss what the findings imply for our understanding of language in general and the English language in particular and what repercussions this has for traditional linguistic standards and models of language teaching.

The results show that English is not a monolithic whole, but that typologically different forms of English are evolving. The Singaporean children make use of features traditionally considered 'wrong' according to British or American English standards. Variationist sociolinguists, however, have long conceptualized these as simply alternative and equally acceptable speech forms. Indeed, the children in the study are all first language speakers of either British English or Singaporean English, a variety of English in its own right. This shows two important and interrelated aspects:

> First of all, languages are changing constantly and this could not be more true for any language than the English language. It has been spread around the globe for the last decades as the result of both British colonization and globalization. Contact with other languages and its appropriation to culturally different contexts have initiated the development of increasingly different speech forms. Maybe, one day, the English language will experience a fragmentation similar to Latin and its offspring Romance languages.

Second, the results suggest that we have to rethink our traditional notion of and orientation towards British and, more recently, American English as the two only valid standards. With new first language varieties of English (Singapore English in our example) being locally and increasingly also internationally accepted as new standards, British and American English might lose their global impact, at least in specific regions in the near future.

25.5 Summary and Outlook

The Singapore study referred to in this chapter is one of the first of its kind that empirically investigates the change from second to first language status in all its detail (Buschfeld 2020). Singapore is indeed leading this development, but other regions such as Malaysia and the Philippines or Cameroon and Botswana are showing increasing numbers of first language speakers of English too. This constitutes a globally increasing research trend in modern linguistics and will have an important impact on the future development and forms of the English language. What seems to be clear is that the English language is no longer exclusively determined and owned by the British and the Americans.

25.6 Further Reading

For demographic information on Singapore see the CIA World Factbook (2020): www.cia.gov/library/publications/the-world-factbook/geos/sn.html and the Department of Statistics Singapore (2020): Census of Population 2020. Statistical Release 1: Demographic Characteristics, Education, Language and Religion; https://www.singstat.gov.sg/publications/reference/cop2020/cop2020-srl (28 May, 2023).

For a discussion of the methodology and more detailed findings of the Singapore study see: Buschfeld, Sarah (2020): "Children's English in Singapore: Acquisition, Properties, and Use." London: Routledge.

For an overview of the developments in the World Englishes research paradigm, see: Buschfeld, Sarah & Kautzsch, Alexander (2020): "Theoretical models of English as a world language," in: Daniel Schreier, Marianne Hundt, and Edgar W. Schneider (eds.), "Cambridge Handbook of World Englishes." Cambridge: Cambridge University Press, 51–71.

For current statistical methods used in Linguistics, see, for example: Levshina, Natalia (2015): "How to Do Linguistics with R: Data Exploration and Statistical Analysis." Amsterdam: John Benjamins; Tagliamonte, Sali, and Baayen, Harald (2012): "Models, forests, and trees of York English: *Was/were* variation as a case study for statistical practice." Language Variation and Change 24, 135–178.

For the software R see R Development Core Team (2014): http://www.R-project.org/, for the R-package 'party' see https://cran.r-project.org/web/packages/party/party.pdf. Based on the ideas for the statistical analyses in this chapter, the authors of this chapter developed the R-package 'PrInDT' (Prediction and Interpretation for Decision Trees); for a documentation see https://cran.r-project.org/web/packages/PrInDT/PrInDT.pdf.

For a discussion of undersampling see G. M. Weiss (2004): "Mining with rarity: A unifying framework," ACM SIGKDD Explorations 6, 7–19.

Chapter 26
Linguistic Manifestations of Cultural Differences Across National Varieties of English: A Methodological Survey

Edgar W. Schneider

Abstract During the colonial and postcolonial periods the English language has been diversified to form many new, stable national varieties in many countries, notably in Asia and Africa. These new 'Englishes' are used in and express a wide range of different world regions, and consequently cultures. A recent linguistic research project, inspired by the sociological discipline of 'cross-cultural analysis', investigated whether these varying cultures find formal manifestations in texts representing different national varieties of English. The present paper focuses on methodological issues of operationalizing this research question, and expounds statistical procedures of testing frequency differences. These goals are pursued on two increasingly abstract levels of linguistic expression, namely terms representative of 'dimensions of culture' and abstract syntactic pattern schemes which may be supposed to be motivated by differences in cultural perspectives.

26.1 Introduction

The colonial period, notably the British Empire, transported and re-rooted the English language in many different corners of the world, and during the postcolonial period (roughly after the mid-twentieth century) globalization has helped to spread it and retain and strengthen its role as the world's leading tool of transcultural communication. In many different historical phases and countries, typically in contact with indigenous languages, new, complex, and stable national varieties of English have emerged. Consequently, a branch of linguistics known as 'World Englishes' has grown to investigate varieties such as Indian, Singaporean, or New Zealand English, typically in comparison to British English, their parent and donor variety. Research on World Englishes has focused on structural properties of these varieties on the levels of pronunciation, vocabulary and grammar, and on similarities

E. W. Schneider (✉)
University of Regensburg, English Linguistics, Regensburg, Germany
e-mail: edgar.schneider@sprachlit.uni-regensburg.de

© The Author(s), under exclusive license to Springer-Verlag GmbH, DE, part of Springer Nature 2024
C. Weihs et al. (eds.), *Statistics Today*, Society, Environment and Statistics, https://doi.org/10.1007/978-3-662-68907-3_26

225

across historical evolutionary processes. With very few exceptions, however, it has not considered the fact that these varieties have all grown in widely different cultural contexts or asked whether fundamental cultural differences, for instance those between 'eastern'/Asian and 'western'/Euro-American cultures, have possibly caused emerging linguistic differences.

The present paper derives from a project which aimed to identify linguistic manifestations of differences between cultures in which English is regularly spoken. Its underlying baseline is a 'platform paper' published in 2021 in a volume on Englishes in the Indo-Pacific region, entitled "Reflections of cultures in corpus texts: Focus on the Indo-Pacific region" (see Further Reading). This paper explicitly attempted to identify linguistic traces of cultural differences in text corpora (for a more detailed treatment of corpora, see Sect. 26.2) representing national varieties of English.

The notion of 'culture' is a fashionable buzzword in the humanities, but one which is notoriously difficult to define and delimitate. The project under scrutiny was inspired and influenced by theories of intercultural communication in anthropology, and in particular by the socio-psychological discipline of 'cross-cultural analysis' as developed by the Dutch sociologist Geert Hofstede. Hofstede posited that coherent community mindsets can be aligned across some systematic 'dimensions of culture', such as collectivism vs. individualism (i.e. whether one's actions are determined more strongly by the interests of the community or one's individual self), or monochromic vs. polychromic time orientations (preferring to perform tasks one after the other or simultaneously). His theory has been immensely influential in the social sciences, has been referred to widely and has triggered literally thousands of follow-up studies in many disciplines (but not in linguistics).

The core question in the present paper is how the theoretical notion of cultural differences (or 'dimensions') is reflected in varying linguistic forms in different countries, and how the quest for such differences can be reliably operationalized. The main formal manifestations of languages and language varieties accessible for systematic and (semi-)automatic screening are texts, obviously, ideally large-scale electronically readable text collections which can be taken to be representative of some linguistic 'universe', such as a national language variety. In linguistics, such large text collections, known as 'corpora', have been compiled and scrutinized over the last few decades, yielding another young sub-discipline, known as 'corpus linguistics'. Such text collections can be automatically searched for pre-defined linguistic target forms (e.g. sequences of specific letters and/or words), trivially by word-processing software but much more effectively by customized software which allows for many other useful types of presentation and analysis, such as word frequency lists, 'key-word-in-context' (KWIC) representations, or automatically calculated word association measurements ('mutual information' design). Currently the best-known and most widely used program for such linguistic purposes is AntConc,[1] a powerful piece of freeware for linguists.

---

[1] cf. https://www.laurenceanthony.net/software/antconc/.

The next important question is which forms to search for, or, more generally, how to identify and justify search forms which manifest a given linguistic issue, in our case linguistic manifestations of cultural dimensions. Corpus searches for specific forms will yield different frequencies of occurrence of any form in several corpora, and the question then, to be submitted to statistical significance testing, is whether or which frequency differences observed are indicative of important factual differences, i.e. whether the varieties compared really behave differently.

It is thus necessary to decide on and define sets of forms which reflect different cultural orientations, search for these forms in electronic text collections (corpora) representing the target varieties in question, count their frequencies in different national corpora, and test frequency differences observed for statistical significance. Consequently, the following methodological issues need to be tackled and will be focused on in this paper:

1. corpora (the choice of text collections considered representative of national varieties of English, including an assessment of their degree of representativeness),
2. linguistic forms considered representative of cultural orientations, and
3. statistical testing procedures regarded as suitable to single out significant differences.

26.2   Electronic Corpora: Representative Text Collections?

In English corpus linguistics, over the last few decades a wide range of electronic text collections have been compiled and mostly made publicly accessible for research purposes; and many of these have targeted variety differences. In fact, the very first of these corpora, compiled as early as in the 1960s and known as the 'Brown Corpus' after its place of compilation (Brown University), explicitly meant to represent (written) American English, and it was followed by an equally-structured British English equivalent a decade later (known as 'LOB', for Lancaster, Oslo, and Bergen, the universities involved), and also an Indian one, the 'Kolhapur' corpus.

Obviously, a core methodological problem for corpus compilers has always been where to get one's texts from and how to select and put them together so that they can be considered and accepted as 'representative' of the language form they are meant to sample (and, in fact, consequently also what exactly the 'universe' to be represented is defined like). The Brown Corpus, for instance, was intended to facilitate objective investigations of properties of 'American English', and since at that time it was inconceivable to collect a sufficiently large body of speech, this was immediately limited to 'written American English'. The other core design properties of this corpus were simply decided on and argued for by the project directors, Henry Kučera and W. Nelson Francis, implying a balance between practical feasibility and representativeness needs: The target magnitude of the corpus is one million words (a so-called 'megaword' corpus), consisting of a fairly large number (500) of relatively short (2000 words, and then up to the end of a sentence) text samples to guarantee versatility and avoidance of skewing by individual overly long texts. The microcosm of 'written American English' was then subdivided into 15 different genres (text types), and a specific number of texts to be selected out of the sum total of 500 was assigned to each of them by the corpus designers, presumably intuitively proportional to the 'importance' or frequency of the respective text types. So, for example, there are 48 'Reportage', 80 'Learned', 29 'Romance and Love Story', and 17 'Religion' texts in the final corpus. Certainly, in terms of achieving representativeness, these are 'educated guesses' or assignments at best—though to my knowledge this particular aspect has never been questioned. The basic principle of corpus compilation thus was established as predetermining text types on the basis of their importance and availability, deciding on target magnitudes, and then feeding select sample texts into the database. This principle has been retained in all later corpus compilation projects, though details have changed drastically.

In recent years, technological advances, in particular the availability of automatic scraping of texts from the web, have removed the limitations imposed by human handling and decision-making procedures and have consequently multiplied available corpus magnitudes. While three decades ago megaword (one-million word) corpora were considered huge, recent collections, often associated with the work and impact of Mark Davies of Brigham Young University,[2] easily reach many billions of words, and these in turn are dwarfed by the Google books collection. For World Englishes research, the Global Web-based English (GloWbE) corpus, with 1.9 billion words from 20 countries, has become a standard tool; similarly, NOW ('News on the Web', both from Davies's above collection) keeps feeding newspaper texts, also from 20 countries, into its collection on a daily basis; currently its size is 15.4 billion words. There is a serious down side to this, however, namely the issue of representativeness, since very many important text and speech act types simply are not represented in written form on the web. GloWbE, for instance, consists of newspaper and blog texts, disregarding other types of utterances and text production, and the question is whether these text types can be considered typical of 'English' in general. So, in these humongous collections, convenience and size reign over deliberate composition; 'big data' thinking has repressed the goal of careful composition.

In the current project, as is the case in very many corpus-based studies in World Englishes today, components of the ICE ('International Corpus of English') project, launched in the 1990s, have been employed. This project's explicit intention was to provide equally structured large text collections representing national varieties of English, for comparison; by now close to twenty such corpora have been made available or are underway.[3] ICE corpora, all compiled by different teams, also have a target magnitude of one million words (in reality they all are somewhat larger), subdivided by a finely-graded roster of text types and featuring 60% of spoken text

---

[2] See his collections at www.English-corpora.org.

[3] See www.ice-corpora.uzh.ch/en.

Table 26.1 ICE corpora
employed and their sizes

Region (corpus abbreviation)	Size (token word count)
Great Britain (GB)	1,036,649
New Zealand (NZ)	1,103,204
Hong Kong (HK)	1,336,831
Singapore (Sing)	1,094,614
India (Ind)	1,097,785

transcripts—a major methodological innovation. They include samples of many different genres, including, for example, direct conversations, class lessons, parliamentary debates, business transactions, demonstrations, broadcast talks, student essays, learned and popular texts covering a range of disciplines, editorials, novels, and many more—a rather diversified representation of human linguistic activities. ICE corpora are thus much more representative of real-life speech performance than, say, the Brown family corpora or GloWbE. Still, the precise corpus composition, the set and relative size of stylistic components and text types, had to be determined and argued for by linguists and project designers. In reality, some limitations of comparability have been observed and have to be accepted. Ideally, all ICE corpora should be equally structured, but in some cases there are practical limitations, resulting from different issues (e.g. the fact that not all required text types are performed in English in the respective, multilingual countries, or that some teams, notably the compilers of ICE-East Africa and ICE-Nigeria, have decided to structure their corpora differently, disregarding the prescribed design).

Interestingly enough, this affects also the central and seemingly trivial problem of determining corpus sizes exactly. Many of the ICE corpora come with a manual, but most of them do not give an exact number of words that the corpus consists of. Software yields a word count number, of course, but in practice different programs, both word processors and customized ones, yield remarkably different numbers, depending on, for instance, whether punctuation and clitics, like *'ll* in *we'll*, or extra-corpus material (e.g. interviewer questions, foreign language material, corpus tags) are counted or not and whether markup errors (found in most corpora, e.g. opened but not closed SGML tags (Standard Generalized Markup Language)[4]) have an impact. Table 26.1 identifies the ICE corpora employed and the corpus size figures used in the present project. For ICE-New Zealand the table represents my sum total of the individual corpus section sizes given in the manual; for all others the table is based on the 'token used for word list' tool by the program WordSmith[5] (which in preliminary comparisons has been assessed to be the most reliable figure).

---

[4] cf. https://www.w3.org/TR/WD-html40-970708/intro/sgmltut.html.

[5] cf. https://lexically.net/wordsmith/.

26.3 Linguistic Forms Representing Cultural Orientations

The items to be automatically counted in these corpora are forms (i.e. precisely defined character strings) which can be argued to represent specific cultural dimensions fairly clearly; I call them 'indicator terms'. As such, they need to be selected and justified by the researcher. In addition to the core quality of serving as reliable cultural indicators, these terms need to avoid some potentially distorting linguistic issues and traps. Ambiguous forms must be avoided. These include mainly homonyms and homographs, that is lexemes which happen to be spelled identically to different, non-target words (e.g. the verb *share* denoting 'cooperative behavior' and thus indicating a collectivist orientation and the noun *share* meaning 'partial ownership', with no such connotation). Another ambiguous form type is polysemic terms, which have a range of different meanings only one of which is relevant to the issue at hand (for example, the adjective *naked* represents a proposed dimension of 'sociosexuality', but not so in *naked truth*). Conversely, test items should be lemmatized, with search forms including all possible variant forms (i.e. with different inflection endings, variant spellings, or variant word classes, so for example the lexeme *protect* should also include the forms *protects, protected, protection*, etc.).

Linguistic representations of culture occur on three levels of increasing abstractness, called 'nexus' in the abovementioned platform paper. Nexus 1 denotes cultural terms and objects (such as food, clothes, social roles, etc., e.g. *kauri, dim sum*, or *sari*). These represent cultures in a simple and straightforward fashion and can be counted easily, so they do not raise any methodologically interesting issues and are therefore disregarded here. Nexus 2 concerns dimensions of culture after Hofstede, so the relationship between searchable forms and the somewhat abstract target concepts is more indirect and in need of discussion and justification. Nexus 3 relates to structural schemes and thus inquires into possible cultural motivations for preferences between alternative grammatical patterns. From the perspective of language organization this would be the most interesting, though also the most abstract type of impact to be unearthed. For each of the two nexus levels 'cultural dimension' and 'structural schemes', I discuss one sample set of distributions.

26.3.1 Example of a Cultural Dimension: Collectivism

Hofstede's most central and well-established cultural dimension is the one distinguishing collectivism versus individualism in a society and its members. It is associated normally with eastern/Asian versus western/Euro-American cultures, respectively, and an interesting cultural-anthropological explanation for its emergence has been suggested: Asia's pastoralism mostly relies on the cultivation of rice, which requires complex irrigation systems, which can only be achieved in labor-intensive communal systems reconciling and downplaying potential conflicts, a community-oriented attitude enshrined in Confucius' widely influential philosophy. In contrast, European hunter-gatherers had to compete for scarce land, so an individual's strength, independence and competitiveness may have constituted a decisive advantage for survival and success—ultimately leading to an appreciation of individualism in societies derived from these contexts.

Which indicator terms may represent such vague social attitudes? There is no principled procedure which mechanically yields such terms; they have to be selected by the researcher, based on some references in the literature and also on personal knowledge and assessment. In the present context, I decided on the following eight terms as indicative of a collective cultural orientation: *tak*/took care of, protect*, loyal*, harmon*, sharing/-ed, together, concerned about*, and *sensitive*. Note that based on AntConc's wildcard search conventions an asterisk ('*') at the end of a lexical item stands for zero or more characters and thus represents alternative endings (so *tak** includes *take, takes, taking*, and *harmon** includes *harmony, harmonize, harmonic*), and a slash ('/') implies alternatives. On this purely formal basis raw frequencies of occurrence of the indicator terms per corpus can then be determined. For instance, *harmon** yields 11 occurrences in the GB corpus, 19 in NZ, 57 in HK, 43 in Sing, and 36 in Ind (for all figures in tabular form see Table 26.2). Obviously, there are frequency differences, and it seems clear that harmony-related words occur more commonly in the Asian corpora compared to the culturally 'western' ones. The obvious important question then, to be addressed in Sect. 26.5, is whether these differences are manifestations of random variability or indicative of some possible causal relation.

Table 26.2 Collectivism-oriented indicator terms—token frequencies + significance levels (by lexeme type, for variety in question versus GB)

Lexemes (lemmatized)	GB	NZ	Hong Kong	Singapore	India
tak*/took care of	9	7	70***	54***	79***
protect*	118	210	215*	186**	147
loyal*	19	28	27	13	28
harmon*	11	19	57***	43***	36**
sharing/-ed	49	78	77	131***	42
together	270	321	471***	350	259
concerned about	19	43	26	22	13
sensitive	3	41***	45***	29***	51***
sum	498	747***	988***	828***	655***

26.3.2   Example of a Structural Scheme: Recipientless Constructions

The research question underlying nexus 3 is: Is it conceivable that cultural differences have produced specific, different grammatical structures, and if so, can such
a motivation be teased out on the basis of numerical distributions? Obviously, this
is a rather abstract question, but linguistically a challenging and interesting one, as
it might offer fundamental insights into one possible determinant, a deeply rooted
cause of language change and emerging language structure. In the discipline, and in
the field of World Englishes, occasional and tentative suggestions along such lines
have been made, but essentially this is largely uncharted territory.

How could such abstract distinctions be captured? Obviously, again, a mechanical and objective search instruction is required. Comparable to indicator terms,
select 'indicator constructions' need to be defined. Since in these cases the
distinction under investigation is rather fuzzy and presumably not visibly manifested
in individual structures, it seems recommendable to devise simple test constructions
which pinpoint the choice in question and neutralize as many other, potentially
interfering factors as possible. Construction choices are often dependent on specific
lexical items, especially verbs (which syntactically determine the number and types
of complements they allow or require), so it makes sense to focus on constructions
associated with specific verbs—ideally high-frequency ones, to facilitate reliable
statistical analysis.

In the present study, I focus on constructions which potentially manifest the same
cultural dimension topic discussed in the previous section, individualism or collectivism. Many verbal constructions encode processes which affect individuals; and if
there are syntactic alternatives as to how to express the form representing a person
in such a construction these might serve as test cases for the culture-construction
hypothesis. In other words, it is conceivable (and has to be rendered testable)
that references to persons in individualist cultures are syntactically highlighted and
focused on (by being placed in a prominent slot, e.g. in syntactic end-focus position)
and alternatively downplayed in collectivist cultures (by avoiding such positions or
avoiding reference to individuals altogether).

For the present discussion, I have selected a simple pattern which I call
'recipientless constructions'. Many English verbs syntactically require a direct
object which typically expresses the individual affected by the activity. This is
the case, for example, with the verbs *assure* and *inform*, where the object needs
to be a person who receives some information (e.g. *I informed/assured her that . . .*).
However, some varieties of English have developed a construction which leaves out
the object (e.g. *I informed that . . .*). In this pattern an individual, the information-
recipient mandatory in British English, is syntactically deleted, hence downplayed
(even disregarded completely)—which may be seen as (and motivated by) a step
towards a non-individualist orientation and information selection.

The recipientless construction *inform*/assur* that . . .* has been searched for
in the ICE corpora. Token frequencies are fairly low (see Table 26.3), but the

Table 26.3 Recipientless constructions—token frequencies + significance level (for variety in question versus GB)

Form	GB	NZ	Hong Kong	Singapore	India
inform/assure that	0	0	4	4	16***

distribution seems straightforward, along the line of the hypothesis of recipientless constructions correlating with collectivist cultures. This pattern is found not at all in culturally western, first-language English-speaking countries (GB and NZ) but it does occur in all Asian second-language speaking countries: 4 times each in HK and Sing, and 16 times, notably more commonly, in India. Again, the question is whether these differences are products of chance variability or can be shown to be significant.

26.4 Statistical Testing

All natural systems, and notably very strongly language, are characterized by some degree of random variability of frequencies and measurements, often distributed along a Gaussian bell curve, and this is the case in the present project as well. The considerations, procedures and counts described in the previous sections consequently yield a two-dimensional contingency table based on nominal data, i.e. the frequencies of occurrence of specific representative forms by varieties. The goal of a statistical assessment then is to test and possibly reject the null hypothesis which posits that there is no intrinsic relationship between form and variety, that the numbers are randomly distributed.

Rather than assessing a complete distribution of the entire data set, we are interested in assessing the frequencies of occurrence of specific forms (indicator terms) found in individual varieties by means of pairwise comparisons, i.e. series of two-by-two matrices involving the frequencies of a target form in two varieties selected for comparison. In the following, only comparisons between individual postcolonial varieties and British English will be reported, since those are the only ones which can be interpreted meaningfully: For all the Englishes under investigation British English is the ancestral, mother and donor variety, so any possible impact of indigenous cultural environments in former colonies should manifest itself by comparing that variety to the GB corpus.

Raw token frequencies found in two varieties should not be directly compared, in particular since the corpora are not really equal in size—this difference has to be factored in as well. The rationale is that in a corpus of a given size every word is a potential site of occurrence of a target form, hence to bring in the effect of varying corpus sizes, for each term its frequency of occurrence (n) has to be juxtaposed to its frequency of non-occurrence (= corpus size − n). Essentially this adopts a procedure suggested and documented by Paul Rayson.[6]

Since some token frequencies in cell entries are very small, the choice of the chi-square test is ruled out, since Cochran's restriction for cell entry sizes (at least 5) is violated. For testing distributions with small token numbers, a log likelihood test (employed by Rayson) or Fisher's Exact test are recommended. In the present study Fisher's Exact was employed, with Bonferroni correction and automatic adjustment of benchmark p values when several pairwise comparisons are conducted.[7,8] In pairwise comparisons, it submitted the number of target form occurrences (n) and the number of non-occurrences (corpus size − n) in the variety in question and in GB, respectively, to a Fisher's Exact test. In the tables below, significance levels returned are symbolized in a conventional fashion by asterisks: $p < 0.001 = ***$, $p < 0.01 = **$, and $p < 0.05 = *$, for p-values p.

26.5 Sample Results

The results obtained by the procedures described above confirm some aspects of the hypothesis and leave others unsupported. As always, they require a closer look and some consideration and qualification. Needless to say, they are meant to be seen as exemplary and illustrative, with a focus on the methodological aspects introduced and discussed.

Based on the data in Table 26.2, overall the hypothesis is strongly confirmed: All three Asian postcolonial varieties, in HK, Sing, and Ind, employ collectivism-oriented indicator terms (see the sum total line) highly significantly more frequently than GB (and NZ does as well), signaling a difference in cultural orientation. With some individual items this distributional pattern of high values across Asia and low ones in GB applies completely as well; cases in point are the word *sensitive*, which is very rarely used in GB, and variants of *harmony*. Other item distributions show the same pattern in a slightly weaker form (e.g. the variants of *take care of*, where NZ sides with GB but the Asian varieties use the predication much more frequently). Other distributions are less clear-cut, especially when it comes to significance levels. For example, *loyal** and *concerned* show weaker but non-significant differences along the lines expected (with GB values mostly lower). However, even for these cases it is worth noting that the frequency of all forms in GB is lower than in (almost) all Asian varieties (one exception is India repeatedly, where *sharing, together* and *concerned* also tend to be avoided). Furthermore, some

---

[6] cf. http://ucrel.lancs.ac.uk/llwizard.html.

[7] cf. Stefan Th. Gries 2008: Statistik für Sprachwissenschaftler, Vandenhoeck & Ruprecht, 243–244.

[8] The implementation of the testing strategy employed a simple R script written and provided by Thomas Brunner (Catholic University of Eichstätt), for which I am grateful to him.

items behave idiosyncratically and show some strong local preferences—consider *protect** in Sing, and *together* in HK.

For the short sample documentation of nexus 3, Table 26.3 assembles and assesses the results. It highlights three aspects: First, the fact that even very low token frequencies can yield really interesting results; second, the value of even simple qualitative comparisons, given that the collectivist pattern in question is completely missing in GB and NZ while it is consistently present in all Asian varieties; and third, that this construction is particularly frequent and highly significant in India, even much more so than in HK and Sing. This is certainly too small a data set and too narrow an approach to accept the distribution as 'proof' of the hypothesis of cultural differences having an impact on syntactic choices, but the result is sufficiently powerful and consistent to rate it as a remarkable piece of support for the hypothesis, especially given its rather abstract and fuzzy character.

26.6 Summary and Conclusion

This chapter has discussed core methodological issues involved in testing whether cultural differences find consistent formal manifestations in texts representing varieties of English, and it has found and weighed some supportive results along such lines. Clearly, given its rather narrow scope, it is essentially exemplary and exploratory in character. It shows, however, that through some deliberate operationalization and the careful choice of methodological tools and steps even what at first seems to be a rather abstract question and distant relationship can be submitted to empirical scrutiny and testing. I have deliberately emphasized and illustrated the fact that qualitative observations in themselves can also be strongly indicative, interesting and meaningful, though quantitative analysis adds something of additional value. Sophisticated statistical machinery, very popular in linguistics today, certainly constitutes a most valuable tool, but it should not (but sometimes seems in danger of doing so) lose sight of the core task of linguistics, i.e. understanding how language works and which forces shape its character and properties. Qualitative considerations and analysis and quantitative tools have to go hand in hand and to support each other.

26.7 Further Reading

For a highly influential theoretical account of how different postcolonial varieties of English have emerged, see Schneider, Edgar W. (2007): "Postcolonial English." Cambridge: Cambridge University Press. For an introductory survey and a comprehensive documentation of the 'World Englishes' research paradigm, see Schneider, Edgar W. (2020): "English around the World. An Introduction." 2nd ed. Cambridge: Cambridge University Press; and Schreier, Daniel, Marianne Hundt & Edgar W.

Schneider (eds.), "The Cambridge Handbook of World Englishes," Cambridge: Cambridge University Press, respectively.

The field of corpus linguistics is also covered in a wide range of textbooks and handbooks, for instance Douglas Biber and Randi Reppen (2020): "The Cambridge Handbook of English Corpus Linguistics". Cambridge: Cambridge University Press.

For a core text on cross-cultural psychology, see Hofstede, Geert (2001): "Culture's Consequences: Comparing Values, Behaviors, Institutions and Organizations across Nations." 2nd ed. Thousand Oaks, CA: Sage.

For broader coverage of the project reported here, see Schneider, Edgar W. (2021): "Platform paper: Reflections of cultures in corpus texts: Focus on the Indo-Pacific region." In Pam Peters and Kate Burridge (eds.) "Exploring the Ecology of World Englishes: Language, Society and Culture." Edinburgh: Edinburgh University Press, 15–45.

Part VII
From Here to Where?

Chapter 27
Is Data Science More Than Statistics? The Bigger Picture

Claus Weihs and Katja Ickstadt

Abstract Beyond statistics, data is also analyzed in data mining and by methods from machine learning and artificial intelligence. In addition, data science is currently propagated as a new discipline. What is the role of statistics then in this diversity of methods and disciplines? We believe that statistics is the most important discipline for the adequate design of the systematic collection of data and for the finding of structures in the data to gain deeper insights from them, in particular concerning the analysis and quantification of uncertainty.

27.1 Data Science: What Is It Anyway?

The present book has showcased many current statistical procedures for the collection and analysis of data. This concluding chapter summarizes the methods introduced in the book and discusses them within the larger context of the research field of data science.

Beyond the field of statistics, data is also analyzed in, for example, computer science. Subdisciplines that belong to this area are data mining, machine learning, and artificial intelligence. Would it not be reasonable to bring together these disciplines and their approaches so that they can mutually profit from each other? In fact, a new discipline has recently been propagated that is dedicated to such thinking: Data science.

But let us start from the beginning: Nowadays, we are exposed to infinite wisdom about data, it seems: "Data is the oil of the 21st century;" or: "We live in the age of Big Data, which enables us to understand business and societal problems that we have not been able to address until now." All this seems to suggest that we need a new discipline, the discipline of Data science. A number of definitions of the term can be found, as for example the following:

C. Weihs · K. Ickstadt (✉)
TU Dortmund, Department of Statistics, Dortmund, Germany
e-mail: claus.weihs@tu-dortmund.de; ickstadt@statistik.tu-dortmund.de

C. Weihs et al. (eds.), *Statistics Today*, Society, Environment and Statistics,
https://doi.org/10.1007/978-3-662-68907-3_27

239

Data science consists of statistics, computer science, and data processing together with communication, social assessment, data-based management, domain knowledge and data-oriented thinking (Cao 2017).

In 1996, the term Data science was part of the title of a statistics conference (International Federation of Classification Societies (IFCS) "Data science, classification, and related methods") for the first time. Even though the term was introduced by statisticians, computer science and business applications seem to play much more important roles in public perceptions of data science, in particular in the era of Big Data, i.e. in times that data has become easily available in great amounts. In fact, data science has a long tradition when it comes to methods of data collection and analysis, especially in statistics; it has simply not been called this way until quite recently.

Originally, the term statistics was used for the description of the data of a state, a country (hence the name statistics). Later, statistics research had temporarily shifted to the mathematical aspects of analytical methods. Already in the 1970s, however, the ideas of John Tukey changed the purely mathematical point of view, e.g., on statistical testing, to deriving hypotheses from data (exploratory point of view), i.e., trying to understand the data before hypothesizing. Therefore, modern statistics has returned to being a data science in the strict sense of the word.

Concerning the connection between statistics and data science, we would like to mention two events. As early as 1997, a radical proposal was made to rename statistics to data science, and in 2015, a number of ASA (American Statistical Association) leaders released a statement about the role of statistics in data science, claiming that "statistics and machine learning play a central role in data science." In our view, statistical methods are crucial in the most fundamental steps of data science. Hence, the premise of our contribution is:

Statistics is one of the most important disciplines to provide tools and methods for identifying patterns in data and by doing so provides deeper insights into data. It is further the most important discipline to analyze and quantify uncertainty.

This chapter aims at addressing the major impact statistics has on the most important steps of analysis in data science.

27.2 Data Science: Steps

27.2.1 General Structure

One of the forerunners of Data Science from a methodological perspective is the famous CRISP-DM (Cross Industry Standard Process for Data Mining). CRISP-DM proceeds along six main steps:

Business Understanding, Data Understanding, Data Preparation, Modeling, Evaluation, and Deployment.

Nowadays, procedures like CRISP-DM are fundamental for applied statistics (cf. also Chap. 2). In our view, the main steps in data science have been inspired by CRISP-DM. However, in CRISP-DM, two important steps are missing, i.e. Data Acquisition and Enrichment, since it works exclusively with so-called observational data, which were collected without a predefined collection plan. We therefore propose an expanded model, which also involves steps of analysis not necessarily part of statistics. In the following, we introduce this expanded procedure, according to which Data Science is a sequence of the following steps (see also Fig. 27.1):

PROBLEM UNDERSTANDING, Data Acquisition and Enrichment, DATA STORAGE AND ACCESS, Data Exploration, Data Analysis and Modeling, OPTIMIZATION OF ALGORITHMS, Evaluation: Model Validation and Selection, Representation and Reporting of Results, and BUSINESS DEPLOYMENT OF RESULTS.

In this sequence, topics in small capitals indicate steps where statistics is less involved. The steps DATA STORAGE AND ACCESS and OPTIMIZATION OF ALGORITHMS are mainly dealt with by computer science (green boxes in Fig. 27.1) and the steps PROBLEM UNDERSTANDING and BUSINESS DEPLOYMENT OF RESULTS (brown boxes) are—at least partly—dealt by the users, e.g., the business use of results is organized by the business management.

Usually, these steps are not just conducted once but are iterated in a cyclic loop. The arrows in Fig. 27.1 also indicate that it is a common procedure that adjacent steps alternate with each other several times. This holds especially for the steps 'Data Acquisition and Enrichment', 'Data Exploration', and '(Statistical Data Analysis and) Modeling,' as well as for 'Statistical Data Analysis and Modeling' and 'Evaluation' (Model Validation and Selection).

In the following, we will discuss the role of statistics in those steps, in which it plays a crucial role (red boxes in Fig. 27.1). As announced in Sect. 27.1, we will come full circle and discuss and provide examples from the previous chapters.

27.2.2   Data Acquisition and Enrichment

Statisticians often emphasize that one should generate data in a systematic way. But why? Indeed, reliable estimates of the effects of the characteristics of interest on the target variable can only be obtained if their levels are chosen in a systematic way by means of Design of Experiments (DoE). If this is not taken into account, as in observational studies, it is generally not possible to distinguish the effects the different characteristics have on the target variable, i.e. their magnitude cannot be reliably determined. Therefore, wherever possible, one should follow the procedures postulated by experimental design. When DoE is used, studies are referred to as

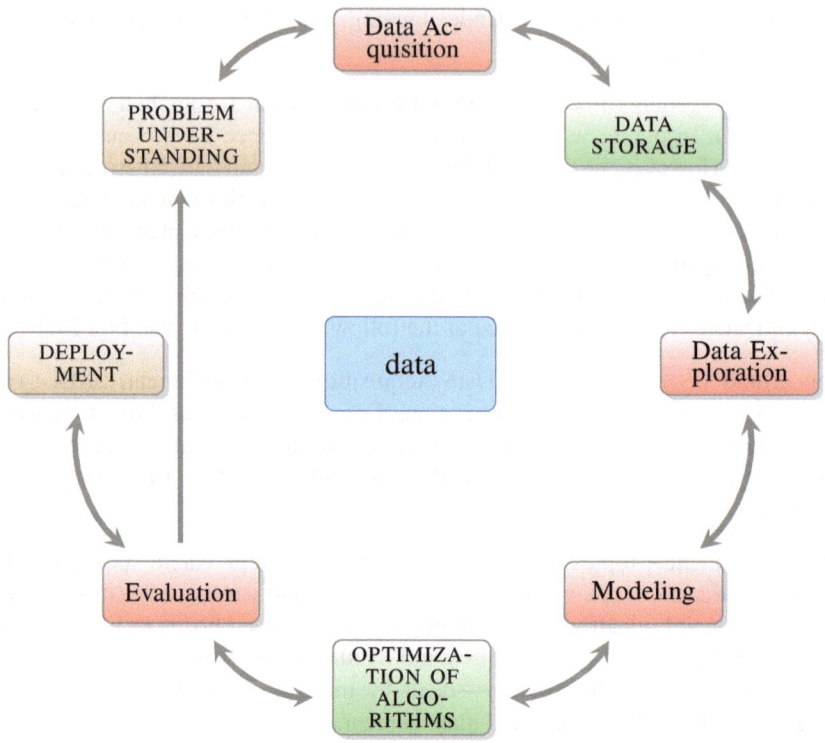

Fig. 27.1 Data Science flow with color coding according to processing discipline: red = statistics, green = computer science, brown = user

controlled or designed. DoE can be used, e.g., to systematically generate new data (data collection) as described in Chap. 3 for determining the optimal number and amount of drug dosages and in Chaps. 18 and 20 in process optimization. Another application area is the selection of those features from a list of potential features that have the greatest impact on the target variable to improve the respective models. Feature selection is used, e.g., in Chap. 10 for instrument recognition and in Chap. 5 for the identification of important genetic features.

Feature selection should guarantee that the selected features are most important not only for the observed sample, but also for other relevant samples. To guarantee this, a method called 'resampling' is typically used, which randomly generates subsets of the observed data. Feature selection then identifies those features which are most important in as many of these 'resamples' as possible. Resampling is also used to generate new samples for testing models; see, e.g., Chap. 10 on the prediction of genres from audio data, Chap. 14 on simulating different realizations of a time series, and Chap. 24 for constructing texts which are similar to the original one.

In addition, data collection is often structured in such a way to allow for the comparison of different conditions in which the data were gathered. One of most important representatives of this type of studies are so-called 'cohort studies', in which two or more groups of, e.g., differently exposed people of one pre-fixed so-called 'cohort' of people are compared concerning their, e.g., risk of disease. The other main study type is the so-called 'case-control study', in which two groups of patients are compared, one diseased (cases), the other not (controls). Both types of studies are discussed in Chap. 6.

Finally, even random data generation follows specific procedures. If a certain distribution is assumed for data which should be generated, this distribution determines the data generation process. Examples for such a task are the uniformly distributed selection of training and test samples in resampling (see above) as well as the selection of lottery numbers as discussed in Chap. 12. Missing data can be generated by so-called imputation methods, which use properties of the observed data to generate the missing values (data enrichment). Such methods are discussed, e.g., in Chaps. 22 and 23.

Statistical methods for data generation and enrichment need to be part of the backbone of data science. The exclusive use of observational data without any control of the levels of the involved variables distinctly diminishes the quality of data analysis and may even lead to the misinterpretation of results. The hope for "The End of Theory: The Data Deluge Makes the Scientific Method Obsolete" (cf. Further reading) appears to be wrong due to noise in the data, no matter how large the data set. Thus, experimental design is crucial for the reliability, validity, and reproducibility of our results.

27.2.3   Data Exploration

Exploratory statistics is essential for data preprocessing to understand the data content. Exploration and visualization of the observed data are the most elaborate parts of data analysis. Examples can be found in particular in Chap. 2 for data preparation, but also in Chap. 1 for visual analysis, Chap. 4 for data smoothing, i.e. the reduction of variation in the data aiming at the elimination of noise and implausible results, again Chap. 4 for the analysis of partial correlations, Chap. 5 for the analysis of correlations between genetic traits, and Chap. 11 for the relation between means and variances.

Data exploration is fundamental to the proper use of analytical methods in data science. The most important contribution from statistics is the notion of 'distribution'. It allows us not only the representation of variability in the data, but also the modeling of information about certain parameters which is known prior to analysis. This kind of information is called 'a-priori information' and is the basic concept of so-called 'Bayesian statistics'; cf., e.g., Chap. 9. Distributions also allow us to choose analytical models and methods appropriately. Concrete examples for the discussion of appropriate distributions can be found, e.g., in Chaps. 7, 9, 14, 15, 17, 18, and 19.

Neglecting distributions in data exploration and modeling makes it impossible to capture the uncertainty of values like means or model parameters. Only if we consider distributions in the data under observation, we gain sufficient insight into uncertainty regions. The presentation of only one estimated value for the characterization of the distribution of a variable or the value of an unknown model parameter without mentioning the respective range of uncertainty would imply a deceptive certainty of the results. Examples for the use of uncertainty ranges can be found, e.g., in Chap. 9 to identify important influences on the outcome of penalties in soccer, in Chap. 19 for forecasts of wire breaks, and in Chap. 24 for the estimation of the difference of text collections like party programs.

27.2.4   Modeling: Statistical Data Analysis

Statistical methods are particularly important for finding structure in data and determining the values of unknown parameters when the exact relationships are not fully known; only statistics can solve analytical tasks while taking uncertainty into account. In the following, we give an overview of the four most important methodological building blocks of statistical data analysis and their use in the preceding chapters of the present book. These building blocks are: Hypothesis Testing, Classification, Regression, and Time Series Analysis.

(a) **Hypothesis testing** is one of the pillars of statistical analysis. Questions arising for data driven problems can often be translated to hypotheses. Hypotheses are the natural link between theory and statistics. Since statistical hypotheses are related to statistical tests, research questions and the validity of existing theoretical approaches can be tested for the available data. Examples of statistical hypotheses and suitable tests are given in Chap. 14 for identifying errors in models for the risk of an asset, in Chap. 16 for testing stochastic trends in greenhouse gas emissions, and in Chap. 26 for differences in cultural indicators in the English of different countries.

(b) **Classification methods** are standard approaches for finding and predicting groups of data representing different properties of the observed objects or subjects. In the so-called unsupervised case, such groups are initially unknown and are identified from a data set without prior knowledge of any cases of such groups. This is usually called clustering. An example of the importance of clusters can be found in Chap. 14 in the analysis of volatility clusters, i.e. groups of similarly high risk assets. In the so-called supervised case, classification rules are determined on the basis of a data set including a discrete variable (called label) whose values represent the different groups of objects/subjects we want to distinguish. Such classification rules are then used to predict unknown group memberships when only influential variables are available. An example of supervised classification is presented in Chap. 10, which deals with finding of classification rules for the prediction of music genres based on audio data. Other examples of classification problems can be found in Chap. 15 for the estimation of the probabilities of loan defaults and in Chap. 25 for the prediction of linguistic characteristics.

(c) **Regression methods** are the main statistical tool to find global and local relationships between features when the target variable is measured. For different distributional assumptions about the target variable, different regression techniques are applied. If a normality assumption appears to be reasonable, linear regression is the most common method. Examples of regressions can be found in Chap. 9 in modeling the outcome of a penalty, Chaps. 18 and 20 in the evaluation of experimental designs, and in Chap. 23 in the imputation of missing data.

(d) **Time series analysis** is concerned with understanding and predicting characteristics which depend on time. Examples of time series analyses can be found in the monitoring of vital functions in intensive care (Chap. 4), in the temporal development of running performances in soccer (Chap. 8), in the estimation of risk measures (Chap. 14), in the process control in 6-sigma analyses (Chap. 18), and in determining mean learning curves with splines (Chap. 21).

In the age of Big Data, a reconsideration of classical methods in statistics and machine learning appears to be necessary since most of the time the calculation effort of complex analysis methods grows overproportionately with the number of observations n or the number of features p. In the case of Big Data, this means that, since n or p or both are large, calculation times will be too high. This resulted in the comeback of simpler algorithms with low time complexity, i.e. a low amount of time taken by the algorithm as a function of the length of the input. In this book, Big Data analyses can be found, e.g., in Chap. 4 in the narrow monitoring of vital functions, in Chap. 5 in the analysis of tens of thousands to millions of genetic values, in Chap. 24 in text data analysis of political party programs, and in Chap. 26 introducing text corpora.

27.2.5 Evaluation: Model Validation and Selection

Model selection has become more and more important over the last years since the number of classification and regression models proposed in the literature has more and more increased. In cases in which more than one model is proposed for, e.g., prediction, statistical tests for comparing models are helpful to rank the models, e.g., concerning their predictive power. Predictive power is typically assessed by means of resampling methods as introduced in Sect. 27.2.2 for feature selection and testing models. This way, the distribution of predictive power characteristics is studied by artificially varying the sample used to determine the model. Characteristics of such distributions can be used for model selection. An example of artificially splitting the data into training and test data can be found in the music data analysis in Chap. 10. In Chap. 5, the predictive ability of individual features is examined.

27.2.6 Deployment of Results

Visualization to interpret structures found in the data and storing of models in an easy-to-update form are very important tasks in statistical analyses. This is essential for the communication of results and to safeguard the deployment of the analysis. Deployment of results, i.e. the post-processing of results by, e.g., process management, is decisive for the usage of the results of the data analysis. It is the last step in our sequence of steps in data science. For statistics, besides visualization and model storage, the most important post-processing task is to report the uncertainties in the data and their respective models. Such uncertainties can be expressed by quantiles (cf. Chap. 9) or by statistical tests (cf. Chap. 16).

27.3 Conclusion

The importance of statistics in data science cannot be overemphasized. This is especially true for areas such as data acquisition and enrichment and advanced modeling. The statistical models and methods described in this book are fundamental to finding structure in data and gaining deeper insights from data, i.e. for successful data analysis. By appropriately considering uncertainty, statistical thinking and sophisticated statistical methods are well-prepared to avoid drawing the wrong conclusions. This is especially true for the analysis of large and/or complex data sets (Big Data).

27.4 Further Reading

In Sect. 27.1, we discuss the basis of statistical data analysis of John W. Tukey (1977): "Exploratory data analysis" (Pearson) and the data science definitions of Longbing Cao (2017): "Data science: A comprehensive overview," ACM Computing Surveys 50(3), and of Jeff Wu (1997): "Statistics = data science?," http://www2.isye.gatech.edu/~jeffwu/presentations/datascience.pdf.

Section 27.2 presents the different steps of CRISP-DM; cf., e.g., Meta S. Brown (2014): "Data mining for dummies" (John Wiley & Sons). The hope for data analysis without theory comes from Chris Anderson (2008): "The end of theory: The data deluge makes the scientific method obsolete" (Wired Magazine). The data science flow in Fig. 27.1 is an extension of the CRISP-DM process flow in Fig. 2.1 in Chap. 2.

A visual analysis loop similar to the data science flow in Fig. 27.1 is proposed in Geppert, L. N., Ickstadt, K., Karl, F., Münch, J., and Steinbrecher, M. (2022): "Visualising complex data within a data science loop: A spatio-temporal example from football," in: Artificial Intelligence, Big Data and Data Science in Statistics: Challenges and Solutions in Environmetrics, the Natural Sciences and Technology, 301–319, Cham: Springer International Publishing, https://doi.org/10.1007/978-3-031-07155-3_13.

Another look at the bigger picture with a focus on artificial intelligence is provided in Friedrich, S., Antes, G., Behr, S., Binder, H., Brannath, W., Dumpert, F., Ickstadt, K., Kestler, H. A., Lederer, J., Leitgöb, H., Pauly, M., Steland, A., Wilhelm, A., and Friede, T. (2022): "Is there a role for statistics in artificial intelligence?," Advances in Data Analysis and Classification 16, 823–846, https://doi.org/10.1007/s11634-021-00455-6.

A paper, similar to this chapter, but with many additional references and without references to the examples of this book, can be found in Weihs, C., and Ickstadt, K. (2018): "Data science: The impact of statistics," International Journal of Data Science and Analytics 6(3), 189–194.

Index

B
Big Data, 39, 203, 245

C
Classification
  classification rule, 87
  logistic regression, 76
  prediction, 17, 87
  quality, 127, 130
  rules, 17, 244
Cluster analysis, 123, 244
CRISP-DM
  business understanding, 13
  data preparation, 15
  data understanding, 13
  deployment, 18
  evaluation, 18
  modeling, 16
  standard procedure, 12

D
Data
  latent variable, 182
  music data, 85
  nominal, 233
  penalties, 74
  preprocessing, 29, 205
  soccer, 65
  specification limits, 156
  Swiss cases of death, 5
Data analysis methodology
  CRISP-DM, 12

data science, 240
  6-sigma, 154
Data collection
  censored observations, 165
  compensation for non-responses, 198
  data enrichment, 241
  demographic data, 59
  design of experiments, 241
  imputation, 200
  incompletely observed data, 60
  indirect questioning, 191
  linguistic data, 214
  non-response, 197
  random choice, 103
  random processes, 163
  subsampling, 199
  supercentenarians, 60
  text corpora, 227
  text data, 203
  weighting adjustment, 199
Data science
  data acquisition, 241
  data exploration, 243
  deployment, 246
  general structure, 240
  model validation, 245
  modeling, 244
Description / exploration , 243
  analysis of relationships, 40
  Bag-of-Words, 205
  boxplots, 70
  contingency table, 233
  correlation, 31
  frequencies, 233

mean-variance profiles, 98
odds ratio, 47
partial correlation, 32
scatter plot, 135
smoothing, 29
Design of experiments , 216
central composite design, 160, 173
coating, 175
doses: number and levels, 23
extreme value problem, 25
fractional factorial designs, 173
full factorial design, 173
hold out, 217
interaction, 175
optimization, 160
Plackett-Burman, 159
screening, 159
surface roughness, 158
training sample, 216
Distributions
exponential, 164
extreme value, 58, 147
Gumbel, 57
normal, 57, 77, 113, 233
Poisson, 66
sum of normals, 77
Weibull, 166

E
Epidemiology and genetics
gene-environment interactions, 50
genetic epidemiology, 49
genetics, 48
types of studies, 47
urinary bladder cancer, 45
Estimation
balanced accuracy, 216
bootstrap, 120, 207
classification, 216
dispersion, 78
distance between texts, 207
with heavy margins, 120
Latent Dirichlet Allocation, 206
L-moments, 147
least squares, 137
maximum likelihood, 166
mean, 164
median, 149
nonparametric, 119
quantile, 78
robust, 148

standard error, 188
trimmed, 149
undersampling, 216

F
Feature selection, 31, 40, 242
Finance
default ordering, 130
dependencies, 109
expected shortfall, 118
investing, 110
obligations and obligors, 125
rating calibration and refinement, 129
rating industry, 125
risk of banks, 117
value-at-risk, 118
variation in time, 111
Floods
frequency, 143
risk and probabilities, 146
types, 149

G
Greenhouse gas emissions
cointegration, 138
CO_2 equivalents, 133
economic activity, 133
environmental Kuznets curve (EKC), 135

H
Horse race betting
betting payouts, 94
favorite-longshot bias, 95
subjective estimates, 96
Hypothesis tests , 244
Bonferroni correction, 234
chi-square, 47
Fisher Exact test, 234
in linguistics, 233
null hypothesis, 233
statistical, 137
stochastic trend, 138

I
Intelligence and education
indicators, 182
latent ability, 186
learning, 185
Levumi online test system, 182

L

Languages
British English, 229
Hongkong English, 229
Indian English, 229
New Zealand English, 229
Singaporean English, 217, 229
World Englishes, 225
Life expectancy
extreme values, 56
finite, 60
Swiss overall, 4
Swiss with accidents, 5
Swiss with cancer, 5
Swiss with cardiovascular deseases, 5
temperature dependence, 8
U.S. supercentarians, 60
Linguistics
AntConc analysis, 226
corpora, 226
cross-cultural analysis, 226
cultural dimension, 230
data collection, 214
linguistic forms, 230
modeling, 216
past tense, 217
predictors, 215
structural schemes, 232
subject pronouns, 219
Lotto
German Zahlenlotto, 102
optimizing the payout, 103

M

Medicine
clinical tests, 21
disease progression, 41
distribution of drugs in the body, 13
drug effects, 18
genetic decision support, 37
intelligent alarm systems, 29
intensive medical care, 27
minimum effective dose, 26
optimization of phase 2, 22
personalized, 37
pharmacokinetics, 11
pre-clinics, 11
therapies, 38
therapy classes, 13
Model
Bayesian, 78
cointegration, 115, 138
consistency, 137

graphical, 32
inverted U-shaped relationship, 135
item-response, 184
latent ability, 182
lifetime, 166
linear, 137
mixed, 78
prediction, 67, 216
prediction interval, 167
predictive accuracy, 40
risk measure, 118
selection, 30, 245
validation, 18, 34, 120, 245
Music
automatic transcription, 90
genre recognition, 91
history, 83
instrument recognition, 89
onset detection, 89
pitch identification, 87

P

Principal components, 32

R

Regression
for imputation, 200
linear, 137, 159, 160, 175
logistic, 76
methods, 245

S

Sampling
compensation of missings, 198
imputation, 200
resampling, 242
re-weighting, 199
test sample, 91, 243
time window, 85
training sample, 87, 243
weighting, 198
6-sigma analysis
control chart, 160
DMAIC, 154
problem definition, 155
process capability, 160
rattling, 153
Soccer
databases, 65
influences on penalty outcomes, 75

leaderboards for goalkeepers and penalty
    takers, 80
match characteristics, 67
number of goals, 66
penalties, 73
running performance, 67
shots on goal, 71

T
Technical processes
coating, 175
deep drilling, 153
lifetime, 164
parameters, 156
products made of several components, 166
reliability function, 164
wear protection, 171
Text analysis
election programs, 208
preprocessing, 205
text collections, 203
topic-based classification, 206
Time series analysis
cointegration, 115, 138
control chart, 160
forecast, 130, 245
simulation, 120
smoothing, 29
spectrum, 85
splines, 186
thresholds, 28, 90, 147
Topics
data science, 239

deep drilling, 153
distribution of drugs in the body, 11
drug dosage, 21
embarrassing truths, 191
epidemiology, 45
finance, 109
flood statistics, 143
genetics, 37, 45
goals in soccer, 65
greenhouse gas emissions, 133
horse race betting, 93
intelligence and education, 181
intensive care medicine, 27
life expectancy and month of birth, 3
lifetime of technical products, 163
lotto, 101
maximum human life span, 55
missing data, 197
music data analysis, 83
penalties in soccer, 73
personalized medicine, 37
ratings, 125
reduction of rejects, 153
reliability of technical products, 163
risk of banks, 117
stock exchange, 109
tension wires, 166
wear protection, 171
World Englishes, 213, 225

U
Uncovering truths
indirect questioning, 191